New Frontiers in Regional Science: Asian Perspectives

Volume 65

This series is a constellation of works by scholars in the field of regional science and in related disciplines specifically focusing on dynamism in Asia.

Asia is the most dynamic part of the world. Japan, Korea, Taiwan, and Singapore experienced rapid and miracle economic growth in the 1970s. Malaysia, Indonesia, and Thailand followed in the 1980s. China, India, and Vietnam are now rising countries in Asia and are even leading the world economy. Due to their rapid economic development and growth, Asian countries continue to face a variety of urgent issues including regional and institutional unbalanced growth, environmental problems, poverty amidst prosperity, an ageing society, the collapse of the bubble economy, and deflation, among others.

Asian countries are diversified as they have their own cultural, historical, and geographical as well as political conditions. Due to this fact, scholars specializing in regional science as an inter- and multi-discipline have taken leading roles in providing mitigating policy proposals based on robust interdisciplinary analysis of multifaceted regional issues and subjects in Asia. This series not only will present unique research results from Asia that are unfamiliar in other parts of the world because of language barriers, but also will publish advanced research results from those regions that have focused on regional and urban issues in Asia from different perspectives.

The series aims to expand the frontiers of regional science through diffusion of intrinsically developed and advanced modern regional science methodologies in Asia and other areas of the world. Readers will be inspired to realize that regional and urban issues in the world are so vast that their established methodologies still have space for development and refinement, and to understand the importance of the interdisciplinary and multidisciplinary approach that is inherent in regional science for analyzing and resolving urgent regional and urban issues in Asia.

Topics under consideration in this series include the theory of social cost and benefit analysis and criteria of public investments, socio-economic vulnerability against disasters, food security and policy, agro-food systems in China, industrial clustering in Asia, comprehensive management of water environment and resources in a river basin, the international trade bloc and food security, migration and labor market in Asia, land policy and local property tax, Information and Communication Technology planning, consumer "shop-around" movements, and regeneration of downtowns, among others.

Researchers who are interested in publishing their books in this Series should obtain a proposal form from Yoshiro Higano (Editor in Chief, higano@jsrsai.jp) and return the completed form to him.

Yoshihide Sekimoto • Yasuhiro Kawahara
Editors

Utilization of Geospatial Information in Daily Life

Expression and Analysis of Dynamic Life Activity

Springer

Editors
Yoshihide Sekimoto
Center for Spatial Information Science
University of Tokyo
Tokyo, Japan

Yasuhiro Kawahara
Division of Art and Sciences
The Open University of Japan
Chiba, Tokyo, Japan

ISSN 2199-5974 ISSN 2199-5982 (electronic)
New Frontiers in Regional Science: Asian Perspectives
ISBN 978-981-19-6904-1 ISBN 978-981-19-6905-8 (eBook)
https://doi.org/10.1007/978-981-19-6905-8

This Springer imprint is published by the registered company Springer Nature Singapore Pte Ltd.
The registered company address is: 152 Beach Road, #21-01/04 Gateway East, Singapore 189721, Singapore

Preface

Recent developments in measurement and information and communication technologies have led to an expanding range of beneficial uses of geospatial information technology in our daily lives. The main origin of this field was the Geographic Information System (GIS), which was developed in Canada in 1967 for the analysis of cities and land uses, and spread rapidly thereafter. GIS was sometimes regarded as an applied system because of its original purpose. Subsequently, it took on the characteristics of a comprehensive information science that deals with the real world when car navigation systems combined with real-time location information from the Global Positioning System (GPS) were developed in the 1980s. The expansion of various web-based mapping applications since the 2000s is commonly known. In addition, significant progress has been made in technologies in relation to major events such as the 1995 Great Hanshin-Awaji Earthquake, the 2011 Great East Japan Earthquake, and COVID-19 after 2020.

Thus, geospatial information science is in the process of systematization in the midst of further development of information processing technology and expansion of information infrastructure in real space. The technical and utilization aspects are simultaneously evolving and stimulating each other in practice. In general, the technical aspects are based on surveys and numerical engineering to measure the position and state of various objects, geography and geology to manage and express the measured objects in the Earth's coordinates, software engineering, data engineering, operations research for efficient processing of graphic information, and recently, communication engineering in terms of mobile terminals and sensor applications. These technologies are used in a wide range of fields, including the analyses of cities, transportation, disaster prevention, recreation, health and medicine, agriculture, forestry, fisheries, culture, ecology, and topography. The technology is easy and convenient for both consumers and researchers to use in their own field of interest, while simultaneously generating interesting, profound, and unknown possibilities that expand our knowledge in a variety of areas.

In this book, individual themes are provided for the reader to experience the diversity and dynamism of geospatial information. Many examples are presented, from collection methods to utilization of such information, while also discussing issues of utilization and future prospects. This book consists of four parts: "Expression of Geospatial Information," "Mobile Communications and Positioning," "Utilization of Geospatial Information in Daily Life," and "Open Data and Personal Data in the Use of Geospatial Information." "Expression of Geospatial Information" and "Mobile Communications and Positioning" describe basic technical aspects, while "Utilization of Geospatial Information in Daily Life" provides applied aspects in individual fields. Each chapter discusses the technical aspects necessary for the use of geospatial information and presents case studies on the chapter's theme. The "Open Data and Personal Data in the Use of Geospatial Information" section describes the use of open data and personal data in Japan, where trends in various countries are attracting attention. We hope that the contents of this book, which describe the current state of geospatial information in daily life and the fields that are showing rapid expansion, will help readers understand and utilize this technology.

Tokyo, Japan Yoshihide Sekimoto
Chiba, Tokyo, Japan Yasuhiro Kawahara

Contents

1 Fundamentals and Applications of Geospatial Information 1
Yasuhiro Kawahara and Yoshihide Sekimoto
1.1 Geospatial Information 1
1.1.1 Use of Geospatial Information 1
1.1.2 Coordinates of the Earth 2
1.1.3 Geodetic System of Japan 4
1.1.4 Surveying and Positioning 5
1.2 Utilization in Various Fields 5
1.2.1 Utilization in Urban Facilities and Land Management . . . 5
1.2.2 Utilization in Transportation Systems and Moving Objects 6
1.2.3 Use in Crime Prevention 6
1.2.4 Utilization in Health and Medical Care 6
1.2.5 Utilization in Disasters 7
1.2.6 Applications in Agriculture, Forestry, and Marine Management 7
1.2.7 Application in Behavior, Ecology, and Cultural Property Survey 7
Reference 8

Part I Expression of Geospatial Information

2 Base Map Development and Surveying in Japan 11
Toshikazu Seto
2.1 Modern Surveying by National Agencies in Japan 11
2.2 Surveying Latitude, Longitude, and Elevation in Japan 12
2.3 Development of Surveying Technology and Satellites in Japan 13
2.4 GSI Maps: Digitization of Topographic Maps and Web-Based Applications 14

2.5 Web-Based Map Services and Utilization of Old Japanese Maps . . 16
2.6 Conclusions 19
References 19

3 Various Representations and Processing Techniques of Geospatial Information: Appropriate Control of Big Data Behind 21
Yoshihide Sekimoto
3.1 Basic Representation of Geospatial Information 21
3.1.1 Graphics and Attributes 21
3.1.2 Overlaying by Layers 22
3.1.3 Software to Operate Geospatial Information 23
3.2 Processing Technology for Geospatial Information 24
3.2.1 Processing Technology 24
3.2.2 Search Technology 25
3.2.3 Metadata Creation Technologies for the Web 26
3.2.4 Map Tiling Technology for the Web 27
3.3 Representation of Moving Objects Including Time and International Standardization 28
3.3.1 What Is Spatio-Temporal Information? 28
3.3.2 Large-Scale Data Processing Technology 29
3.3.3 Standardization of Geospatial Information 29
3.4 Conclusion 30
References 30

Part II Mobile Communications and Positioning

4 Ground-Based and Space-Based Positioning 33
Yasuhiro Kawahara
4.1 Satellite-Based Positioning 33
4.1.1 History of Radio Navigation 33
4.1.2 GNSS (Global Navigation Satellite System) 34
4.1.3 Corrective Positioning 37
4.1.4 Quasi-Zenith Satellite System 38
4.2 Ground-Based Positioning 38
4.2.1 Types of Wireless Positioning 38
4.2.2 Wireless Positioning Applications 41
4.2.3 Dead Reckoning (xDR) 44
4.2.4 Indoor Positioning 44
References 45

5 Observation from Space: Remote Sensing 47
Masahiko Nagai
5.1 What Is Remote Sensing? 47
5.1.1 Satellite Remote Sensing 47
5.1.2 History of Remote Sensing 48
5.2 Earth Observation Satellite 50

5.2.1 Sensor Type 50
5.2.2 Orbits of Earth Observation Satellites 51
5.2.3 Application Fields of Remote Sensing 52
5.3 New Era of Satellite Remote Sensing 52
5.3.1 New Business Through Remote Sensing Technology . . . 52
5.3.2 Use of Small Satellites 53
5.3.3 Satellite Constellation 53
5.4 Summary 55

Part III Utilization of Geospatial Information in Daily Life

6 Applications in Urban Facility and Land Management: From Urban Facility Management to Smart Cities 59
Yoshihide Sekimoto
6.1 Urban Facility Management Starting from Underground Facilities 59
6.1.1 History Starting from Gas Explosion 59
6.1.2 Road Management Information 60
6.1.3 Data on Various Urban Facilities 60
6.2 To Grasp Buildings and Land 62
6.2.1 Understanding Property Taxes 62
6.2.2 For Transparency and Revitalization of Real Estate Transactions 63
6.2.3 Understanding the Age of Buildings for Disaster Prevention 64
6.3 For Urban Planning 65
6.3.1 Basic Urban Planning Survey 65
6.3.2 For Solving Various Urban Issues 65
6.3.3 Sharing City Data Between Citizens and Officials 67
6.4 Conclusion 68
References 69

7 Applications in Transportation Systems and Mobile Devices: Capturing the Movement of People 71
Yoshihide Sekimoto
7.1 Use in Transportation Systems 71
7.1.1 Use in Road Transportation Systems 71
7.1.2 Use in Railroad Systems 72
7.1.3 Use in Bus Systems 73
7.2 Use in the Flow of People 74
7.2.1 Reproduction of People Flow in Urban Areas Using Person Trip Survey Data 74
7.2.2 Utilization of GPS and Base Station Usage Data of Mobile Terminals 75

7.2.3 The Suica Incident and the Concept of Personal Information . . . 76
7.2.4 Positioning of Location Information . . . 77
7.2.5 For Proactive Use . . . 78
7.3 Future People Mobility and Transportation Systems . . . 78
7.3.1 MaaS System Using Smart Phones . . . 78
7.3.2 Automated Driving and High-Precision Maps . . . 79
7.3.3 People Mobility and Behavior Change . . . 80
7.4 Conclusion . . . 81
References . . . 81

8 Applications in Crime Prevention . . . 83
Ikuho Yamada
8.1 A Brief History of the Use of Maps in Crime Research . . . 83
8.1.1 Crime Mapping . . . 83
8.1.2 Introduction of Computers into Crime Analysis . . . 84
8.1.3 Applications for Safety and Security of Neighborhood Communities . . . 85
8.2 Various Applications of Geospatial Information in Crime Prevention . . . 86
8.2.1 Visualization of Crime Occurrence . . . 86
8.2.2 Environmental Factors Related to Crime Risk . . . 88
8.2.3 Geographic Crime Prediction . . . 90
8.3 Summary . . . 92
References . . . 92

9 Applications in Health and Healthcare . . . 95
Ikuho Yamada
9.1 Spatial Distribution of Diseases . . . 95
9.1.1 Disease Maps . . . 95
9.1.2 Spatial Diffusion of Disease . . . 98
9.1.3 Spatial Surveillance of Disease . . . 99
9.2 Locations of Healthcare Facilities and Services . . . 100
9.2.1 Demand Prediction for Healthcare Facilities and Services . . . 100
9.2.2 Accessibility to Healthcare Facilities and Services . . . 101
9.3 Human Health and Environments . . . 101
9.3.1 Potential Associations Between Human Health and Environments . . . 101
9.3.2 Physical Environments and Health . . . 102
9.3.3 Broader Concept of the Living Environments and Health . . . 103
9.4 Summary . . . 105
References . . . 105

10 Application in Times of Disaster: Through the Great Hanshin Earthquake, the Great East Japan Earthquake, and COVID-19 . . . 107
Yoshihide Sekimoto
10.1 Lessons from the Great Hanshin Earthquake . . . 107
10.1.1 Geographic Information Utilization in Disasters Has Begun in Earnest . . . 107
10.1.2 Government System in Place . . . 108
10.2 The Great East Japan Earthquake . . . 109
10.2.1 Emergency Phase . . . 109
10.2.2 Recovery Phase . . . 110
10.2.3 Preservation of Long-Term Memories . . . 113
10.3 Preparing for COVID-19 and Future Catastrophes . . . 114
10.3.1 Utilization of Human Flow in COVID-19 . . . 114
10.3.2 Improvement of Measurement Technology for Natural Disasters . . . 116
10.3.3 Improvement of Simulation Technology . . . 117
10.4 Conclusion . . . 118
References . . . 118

11 Utilization in Agriculture, Forestry, and Marine Management . . . 119
Masahiko Nagai and Yasuhiro Kawahara
11.1 Agriculture . . . 119
11.1.1 Growth Monitoring . . . 119
11.1.2 Orchards and Golf Courses . . . 121
11.2 Forestry . . . 122
11.2.1 Forest Monitoring . . . 122
11.2.2 Bamboo Forest . . . 122
11.3 Maritimes Management . . . 124
11.3.1 Vessel Monitoring . . . 124
11.3.2 Coral Reefs . . . 125
11.3.3 Fish Group Exploration . . . 125
11.4 Summary . . . 128

12 Utilization in Behavior, Ecology, and Cultural Property Surveys . . . 129
Yasuhiro Kawahara
12.1 Mobile Sensing . . . 129
12.1.1 Mobile Computing . . . 129
12.1.2 Human Sensing . . . 131
12.1.3 Human Probes . . . 132
12.2 Biologging . . . 133
12.2.1 Recording Wildlife Information . . . 133
12.2.2 Data logger . . . 134
12.2.3 Biologging on Land, in the Sea, and Air . . . 135
12.3 Use of Geospatial Information in Cultural Property Surveys . . . 137

12.3.1 Measured Drawings of Archaeological Remains 137
12.3.2 Various Surveying and GIS 138
References 141

13 International Applications 143
Masahiko Nagai
13.1 Global Issues 143
13.1.1 SDGs and Space Technologies 143
13.1.2 Global Warming 144
13.1.3 Climate Change 145
13.1.4 COVID-19 146
13.1.5 Forest Fires 146
13.1.6 Malaria 147
13.2 International Cooperation in Times of Disaster 148
13.2.1 International Disaster Charter 148
13.2.2 Sentinel Asia 150
13.3 Summary 151
References 151

Part IV Open Data and Personal Data in the Use of Geospatial Information

14 The Advent of a Participatory Data Society and the Usage of Open Data 155
Toshikazu Seto
14.1 Data Sharing Through User Participation 155
14.1.1 Crowdsourcing 155
14.1.2 Participatory Mapping 157
14.1.3 Volunteered Geographic Information 158
14.2 Grassroots Sharing and Practice of Geospatial Information 158
14.2.1 OpenStreetMap 158
14.2.2 Crisis Mapping Activity 161
14.2.3 Mapillary: Web-Based Service for Sharing Street Landscape Images 163
14.3 Open Data and Geospatial Information 163
14.3.1 Open Data and Utilization for Geospatial Information . . . 163
14.3.2 Utilization of Data Through Citizen Collaboration for Open Government 164
14.4 Conclusions 165
References 166

15 Future Geospatial Information in Harmony with Advanced Technology and Human Life 167
Yoshihide Sekimoto
15.1 Various Advanced Measurement Technologies 167
15.1.1 Sophistication of Measurement and Control from the Air 167

15.1.2 Advancement of Ground-Based Measurement and Control 168
15.1.3 Measurement of People's Behavior 168
15.2 Diffusion of Artificial Intelligence Technology and Beyond 169
15.2.1 Breakthroughs in Deep Learning and the Importance of Data 169
15.2.2 Pseudo Data Generation (Fake Data) Using GAN 171
15.2.3 Toward a Truly Collaborative Society that Raises the Awareness of Citizens 172
15.3 For the Betterment of Society 173
15.4 Conclusion 173
References 173

About the Editors

Yoshihide Sekimoto is Professor in the Center for Spatial Information Science at the University of Tokyo. He received his PhD in civil engineering from the University of Tokyo in 2002. Previously, he worked as researcher in the National Institute for Land, Infrastructure and Management, MLIT, from 2002 to 2007; Project Lecturer in the Center for Spatial Information Science, University of Tokyo, from 2007 to 2010; Project Associate Professor in the Center for Spatial Information Science, University of Tokyo, from 2010 to 2013; Associate Professor in the Institute of Industrial Science, University of Tokyo, from 2013 to 2020; and Professor in the Center for Spatial Information Science, University of Tokyo, from 2020. He is a Member of ACM, IEEE, Transportation Research, GIS Association of Japan, Japan Society of Civil Engineers, and Japan Society of Traffic Engineers. His laboratory is "Human Centered Urban Informatics." His research interests include large-scale data processing of moving objects and infrastructure data management.

Yasuhiro Kawahara is Associate Professor in the School of Graduate Studies of Arts and Sciences at the Open University of Japan. He majored in Environmental Information and Microsystems in the Graduate School of Frontier Sciences at the University of Tokyo and received a PhD in 2005. Previously, he worked as Project Assistant Professor at the University of Tokyo from 2006 to 2010, Lecturer at Kobe University in 2010, Visiting Associate Professor at Tokyo University of Science from 2011 to 2013, and Associate Professor at the Open University of Japan since 2011. He is visiting researcher at Jean Monnet University from 2019. He is a member of the executive board of Human Communication Group in IEICE (The Institute of Electronics, Information and Communication Engineers in Japan). He published papers on Studies in Mobile Computing and Environmental Cognitive Science. His research concern is to develop methods of monitoring human cognition and behavior and evaluate them under particular environmental stimulus in daily life.

Chapter 1
Fundamentals and Applications of Geospatial Information

Yasuhiro Kawahara and Yoshihide Sekimoto

Abstract Geospatial information is used in various scenarios, from daily life to industry and administrative management, and this chapter explains its definition and the principles necessary for its use. The geodetic system in Japan is also discussed, and its use in surveying and positioning is outlined. Taking into account the needs of current times, matters in the field of utilization of geospatial information covered in this book are briefly introduced, which includes static to dynamic information that changes over time.

Keywords Geospatial information · Coordinate system · Geodetic system · Latitude · Longitude

1.1 Geospatial Information

1.1.1 Use of Geospatial Information

In our daily lives, there are many opportunities to use information linked to locations. Look at a street map to determine where we are. Check the weather map to determine when rain is likely to occur. We use information accompanied by location by referring to a map on which information relevant to our purpose is illustrated; for

---

This book is based on Kawahara and Sekimoto eds. (2022), Basics and utilization of the geospatial information, 2022, Foundation for the Promotion of The Open University of Japan (in Japanese), which has been modified for chapters.

Y. Kawahara (✉)
Division of Art and Sciences, The Open University of Japan, Chiba, Japan
e-mail: kawahara2@ouj.ac.jp

Y. Sekimoto
Center for Spatial Information Science, University of Tokyo, Tokyo, Japan
e-mail: sekimoto@iis.u-tokyo.ac.jp

Y. Sekimoto, Y. Kawahara (eds.), *Utilization of Geospatial Information in Daily Life*, New Frontiers in Regional Science: Asian Perspectives 65,
https://doi.org/10.1007/978-981-19-6905-8_1

example, to determine the location of our destination on a map and check the route to that point. In recent years, with the spread of the Internet and mobile information communication terminals, such as cell phones and smartphones, these activities in daily life are often performed using electronic maps. By simply scrolling through the display of the telecommunication terminal, viewers can seamlessly browse maps of the world and obtain as needed the information embedded in these maps. With the popularization of such electronic maps, maps that can be selected according to the purpose and display the necessary information depending on the situation are now available on a daily basis. Services such as car navigation systems and Google Maps that enable such use have become familiar to the majority of consumers who drive or use the internet. Such location-based information is called geospatial information and is defined in Japan under the Basic Act on the Promotion of Utilization of Geospatial Information, which is a law. This book describes how such information is produced, and presented to and used by users. In this context, many examples are presented, especially those specific to Japan.

1.1.2 Coordinates of the Earth

The position on Earth is expressed in terms of latitude and longitude. The latitude is 0° at the equator and is divided into 90° to the north and south of the equator, and counted as north and south latitudes, respectively. Longitude is determined by taking the prime meridian as 0°, dividing the east and west of this line into 180° each, and denoting east as east longitude and west as west longitude (Fig. 1.1). This system, expressed by latitude and longitude, is called the coordinate system. This section describes the current definition of this coordinate system and its recent utilization in Japan.

The shape of the Earth, expressed as the mean sea level over the entire globe, is called a geoid. A geoid is a surface that is orthogonal to the direction of gravity and is undulating in shape because the density of the crust near the Earth's surface varies from place to place. To set up a coordinate system for determining the latitude and longitude, it is convenient to regard the earth as an ellipsoid, and an earth ellipsoid is defined that approximates the geoid well. Currently, to use the coordinate system universally, the center of this earth ellipsoid is coincident with the center of gravity of the Earth (Fig. 1.2).

The World Geodetic System is based on this earth ellipsoid, with its axes at the equator and prime meridian, from which latitude and longitude are drawn. In other words, a geodetic system is a standard for expressing positions on the Earth using latitude and longitude, and a world geodetic system is one in which the coordinate origin and the center of gravity of the Earth are coincident such that the geodetic system can be used in common worldwide. The reference ellipsoid and coordinate system used are determined in the particular world geodetic system used by each country. The reference ellipsoid used in a global positioning system (GPS), which is

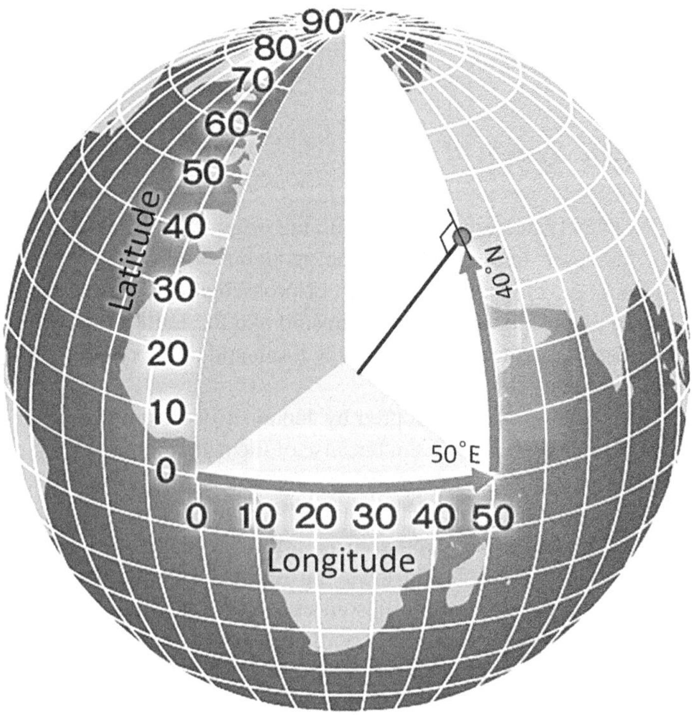

Fig. 1.1 Longitude and latitude

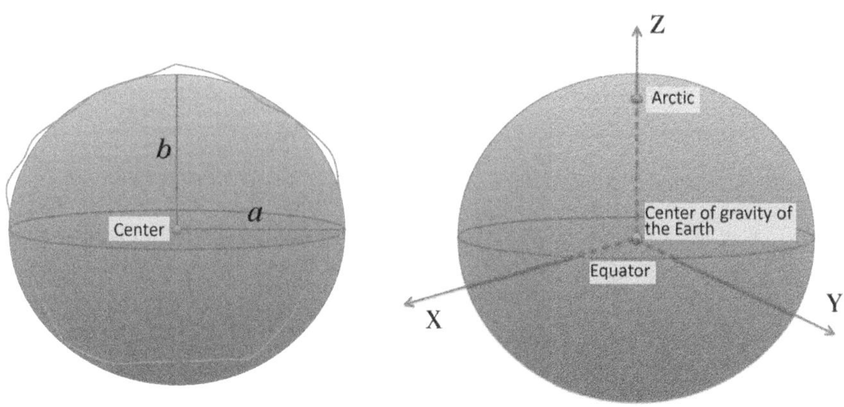

Fig. 1.2 Earth ellipsoid and coordinate system

a satellite positioning system, is called WGS 84, and the coordinate system used is WGS 84.

1.1.3 Geodetic System of Japan

In Japan, Bessel 1841, an earth ellipsoid with the origin of the longitude and latitude set in Japan, was used until 2002, and the geodetic system was called the Japan Geodetic System. An ellipsoid, which is not problematic for domestic surveying, was used. The relationship between the ellipsoid and the Earth is shown in Fig. 1.3, and as can be seen, the center of gravity of the Earth and the center of the ellipsoid do not coincide.

In 2002, the geodetic system adopted by Japan shifted from the Japan Geodetic System to the World Geodetic System because of the need for a more accurate world geodetic system resulting from the development of space geodetic technology and globalization. Since then, surveying conducted by local governments, the creation of maps and map databases for geographical information systems (GIS), and the display of longitude and latitude in laws and ordinances have been based on the World Geodetic System. The earth ellipsoid used in the world geodetic system was GRS 80, and the geocentric Cartesian coordinate system when this system started was ITRF (International Terrestrial Reference Frame) 94. ITRF 2008 was revised because of the large-scale crustal deformation caused by the Great East Japan Earthquake in 2011, and ITRF 2008 was adopted as the coordinate system in the East Japan and Hokuriku regions.

The latitude and longitude expressed in the Japanese geodetic system differ from the latitude and longitude of the same point expressed in the world geodetic system. For example, if a point indicated by the latitude and longitude of the Japanese geodetic system is expressed by the latitude and longitude of the world geodetic

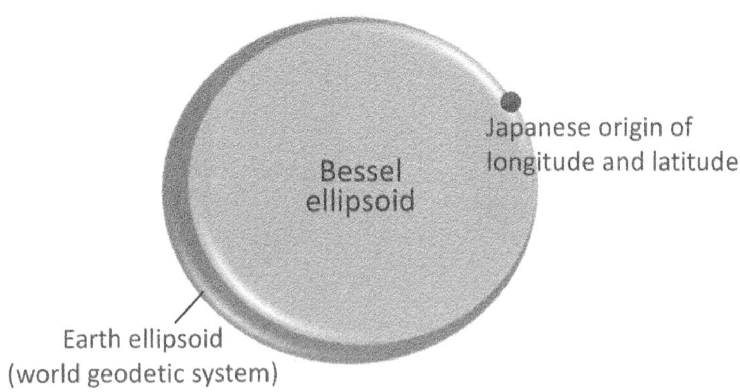

Fig. 1.3 Earth ellipsoid referred by the Japanese geodetic system (used until 2002)

system, the latitude changes by approximately +12s and the longitude changes by approximately −12s around Tokyo. Converting this to a distance, the shift is approximately 450 m northwest. Therefore, unifying the geodetic systems is necessary when comparing locations using maps of the different geodetic systems in Japan.

1.1.4 Surveying and Positioning

To create a map, it is necessary to measure the shape of landforms and artifacts on the Earth. Surveying is the process of measuring the locations of nature and artifacts on Earth and expressing these numerically or as drawings.

The measurement of the position of a stationary or moving object is known as positioning. Positioning technology has evolved through the development of astronomical observation technology and the invention of compasses. A world map was created during the Age of Discovery through navigation and surveying. In the twentieth century, radio-wave positioning technology was used to determine the position of ships and aircraft from land. In the latter half of the twentieth century, satellites for positioning (GNSS, global navigation satellite system) were launched, and using radio signals from multiple satellites, the position of a radio receiver could be estimated with an accuracy of several meters to several tens of meters. The GPS, originally developed for military use in the United States, can determine a three-dimensional location by receiving time and orbital information from multiple satellites and is now available free of charge to anyone in everyday life.

Photogrammetry and remote sensing are technologies used for surveying from satellites and aircraft. Remote sensing is a method of obtaining wide-area locational information using electromagnetic wave measurement devices mounted on aircraft and satellites. It is used in various fields such as agriculture, forestry, fisheries, and administrative planning, including for the estimation of land use and oceanographic conditions.

1.2 Utilization in Various Fields

This section provides an overview of the utilization of geospatial information in the fields covered in this book.

1.2.1 Utilization in Urban Facilities and Land Management

Cities have various components, including roads, underground structures, buildings, and land, and the use of geospatial information is effective in managing these elements and urban planning. For example, in many cities, aerial photographs are used to track the expansion of urbanized land on maps and to manage buildings and land. Chapter 6 describes how to handle geospatial information and obtain a bird's-eye view of a city from the perspectives of urban facility management and urban planning.

1.2.2 Utilization in Transportation Systems and Moving Objects

Many of us have recognized the use of location information from moving objects in daily life through the use of car navigation systems. Geospatial information of moving objects is also used in railroad and bus transportation systems, including automobile transportation systems linked to car navigation systems, to capture the movement of mobile vehicles and people. In addition, the development of mobile information communication terminals equipped with GNSS modules and communication infrastructure has made it possible to visualize user movement. Chapter 7 provides an overview of the technology and its evolution in detecting people's movements using transportation systems and mobile terminal data, as well as the handling of personal information related to location information and how it should be in the future.

1.2.3 Use in Crime Prevention

Crime is concentrated within a limited area and time. Therefore, geospatial information can be effectively used in crime research and prevention. Chapter 8 describes the research and approaches for safety and security using geospatial information, including the visualization of crime occurrence, analysis of environmental factors related to crime, and crime prediction.

1.2.4 Utilization in Health and Medical Care

Geospatial information is also used in the health and medical fields. Disease maps depicting the number of disease outbreaks, combined with population distribution and environmental data, can be analyzed using a GIS (geographic information system) to help prevent disease outbreaks and their spread. In health and medical services, geospatial information is useful for accessibility and demand analysis for the efficient provision of such services. In Chap. 9, approaches using geospatial information are described, focusing on the spatial distribution of diseases, spatial arrangement of health and medical services, and environment surrounding people.

1.2.5 Utilization in Disasters

Geospatial information is effectively used in each phase of a disaster: disaster prediction, emergency response, post-disaster recovery support, and disaster documentation. For example, monitoring the location of mobile information communication terminals during disasters can capture the constantly changing flow of people, which is useful for securing evacuation routes and organizing traffic. In Chap. 10, the history of the Great Hanshin-Awaji Earthquake and the Great East Japan Earthquake in Japan, together with the recent response to COVID-19, is reviewed from the perspective of information technology utilization on the use of geospatial information to comprehensively cope with disasters.

1.2.6 Applications in Agriculture, Forestry, and Marine Management

In agriculture, geospatial information is used in a multifaceted way. Geospatial information is used by governments to survey land use, by producers to manage field use, to monitor crop growth, to monitor the components of crops as they change with growth, to manage the timing and location of fertilizer application, and on large farms to record the location of operations using agricultural equipment and to plan the order of harvesting using farm machinery and efficient facility use. Chapter 11 describes the use of geospatial information in forestry and marine management, in addition to agriculture, with a focus on the use of remote sensing technology.

1.2.7 Application in Behavior, Ecology, and Cultural Property Survey

The miniaturization and power saving of information and communication devices and sensors have made it possible to record behavioral and physiological information combined with geospatial information using devices carried by humans or sensors attached to animals. This information is used for behavioral recognition and ecological research. Furthermore, to estimate the lifestyle and ecology of past eras, archaeological sites are organized as geospatial information in the field of archaeology. Chapter 12 describes the technologies and use of geospatial information to realize these applications, with case examples.

Reference

Kawahara Y, Sekimoto Y (2022) Basics and utilization of the geospatial information. Foundation for the Promotion of The Open University of Japan, Tokyo. (in Japanese)

Part I
Expression of Geospatial Information

Chapter 2
Base Map Development and Surveying in Japan

Toshikazu Seto

Abstract Current geospatial information is maintained by precise latitude and longitude coordinates and altitude on Earth. This chapter mainly explains the status of surveying and map maintenance in Japanese national organizations and draws attention to modern map sharing through the web. Notably, in recent years, many paper maps of the past have been developed and stored as digital archives, and their use in daily life is increasing.

Keywords Geospatial Information Authority of Japan · Modern surveying · Web Map Service · Digital archive · Old Map

2.1 Modern Surveying by National Agencies in Japan

To produce maps with accurate scales, it is necessary to measure the location and area of natural and man-made objects on Earth and represent them with numerical values and drawings (Rooney 2015). In Japan, mapping projects based on modern surveying techniques were carried out on a national scale from approximately 1870 onwards, during the Meiji period. Historically, it commenced in 1869 when the cadastral registration map section was in the office of the Geography Department, Ministry of Civil Service. In 1888, it was reorganized as the Japanese Imperial Land Survey, and finally, in 1945, it became the Geographical Survey Institute (GSI). In April 2010, its name was changed to the "Geospatial Information Authority of Japan."

In 1876, the first long-distance survey was conducted between Tokyo and Shiogama in the Miyagi Prefecture. Colin Alexander McVean, a civil engineer from Scotland, is said to have provided technical guidance at this time, and this is remembered as the time British surveying techniques took root in Japan (Une 2021).

T. Seto (✉)
Faculty of Letters, Komazawa University, Tokyo, Japan
e-mail: tosseto@komazawa-u.ac.jp

Y. Sekimoto, Y. Kawahara (eds.), *Utilization of Geospatial Information in Daily Life*, New Frontiers in Regional Science: Asian Perspectives 65,
https://doi.org/10.1007/978-981-19-6905-8_2

On the other hand, Claude Jourdan introduced French surveying techniques in Japan and produced a 1:20,000 survey map of Tokyo by planimetry in the 1880s. This map was characterized its French-style colored format which grasped not only the topography of the time, but also the vegetation and land use, simultaneously including representative landscapes of the region as landscape paintings around the map. From 1885 to 1912, topographic maps at a scale of 1:20,000 were produced for more than 1500 areas of Japan based on triangulation and leveling techniques; these are the topographic maps (generally at a scale of 1:25,000) produced by the GSI today.

2.2 Surveying Latitude, Longitude, and Elevation in Japan

In Japan, there is one reference point defining the positional coordinates for surveying in the country, the origin of latitude, longitude, and altitude, which serves as the reference point for all surveys. These are called the "Japanese origin of longitude and latitude" and "Japanese datum of leveling." The Japanese origin of longitude and latitude is located at 139°, 44 min, 28.8869 s east longitude and 35°, 39 min, 29 s, 1572 north latitude, and its address is 18-1, Azabudai 2-Chome, Minato-ku, Tokyo (the former site of Tokyo Observatory). It is based on the World Geodetic System, which is a reference system for a position commonly used around the world. Multiple reference locations and triangulation points have been established in Japan by surveying from this point of reference.

The next standard for determining height is the elevation of land, which is measured by setting the mean sea level of Tokyo Bay at 0 m. It was set up in the northern area of the National Diet forecourt in Nagata-Cho, Chiyoda-ku, Tokyo.

The Japanese datum of leveling was built in the style of a Roman temple to prevent fluctuations in height over time and was designated as an important national cultural property in 2019. The elevation was determined to be 24.500 m above sea level in 1891. However, in view of crustal movements caused by the repeated occurrence of major earthquakes in Japan, it was revised to 24.3900 m with the occurrence of the Great East Japan earthquake in 2011. The height of each of the level points located at intervals of about 2 km along major roads in Japan is determined by level surveying based on the origin of the Japanese level; the level points, of which there are about 17,000 in Japan, collectively serve as the standard for determining the precise height of the land in the area. In addition, a network of 1 equal triangulation points covering the whole of Japan at intervals of approximately 40 km was established, and approximately 1000 points were established by 1915.

2.3 Development of Surveying Technology and Satellites in Japan

In the 1970s, light-wave rangefinders were put to practical use. It became possible to perform trilateration to determine the location of triangulation points more accurately by accurately measuring the straight-line distance of each side. In the twenty-first century, global navigation satellite system (GNSS) surveying in Japan has been replaced for more accurate surveying. Currently, GNSS surveying has become the mainstream of geodetic surveying because it enables three-dimensional (3D) high-precision surveying and improves the efficiency of surveying work.

To support this technology, reference points that can continuously receive GNSS signals per second have been established in approximately 1300 control points throughout Japan. Each of these points is called a "GPS-based Control Station," and they are used for many surveying purposes because they can accurately determine the longitude, latitude, and ellipsoid height of any location in Japan. The observation data from the electronic reference points are constantly transmitted to the GSI and are provided to the public through the Internet. For example, when the Great East Japan earthquake occurred, the electronic reference point in Ishinomaki City, Miyagi Prefecture, sank approximately 1.2 m by moving about 5.3 m in the east-southeast direction.

MICHIBIKI, Japan's Quasi-Zenith Satellite System (QZSS), can calculate location information by transmitting radio waves from its satellites, like the global positioning system (GPS) in the United States. MICHIBIKI is a satellite with an orbit that moves in figure-eight pattern with north-south symmetry over Japan and Australia, synchronized with the Earth's rotation, and is therefore suitable for local positioning services (Fig. 2.1). MICHIBIKI has been operated under four satellites from November 2018 onwards and has been planned to be under seven satellites in the financial year ending 2024, with the aim to achieve stable location-based positioning in Japan.

Coordinate values calculated by GNSS surveying are generally expressed in the World Geodetic System 1984 (WGS84) coordinate system, which is the world geodetic system constructed and maintained by the United States. The WGS84 coordinate system and the International Terrestrial Reference Frame (ITRF) coordinate system are almost identical, with the origin coinciding with the center of the earth. The time difference between multiple radio waves used for calculations is very small considering the speed of radio waves flying at 300,000 km/s.

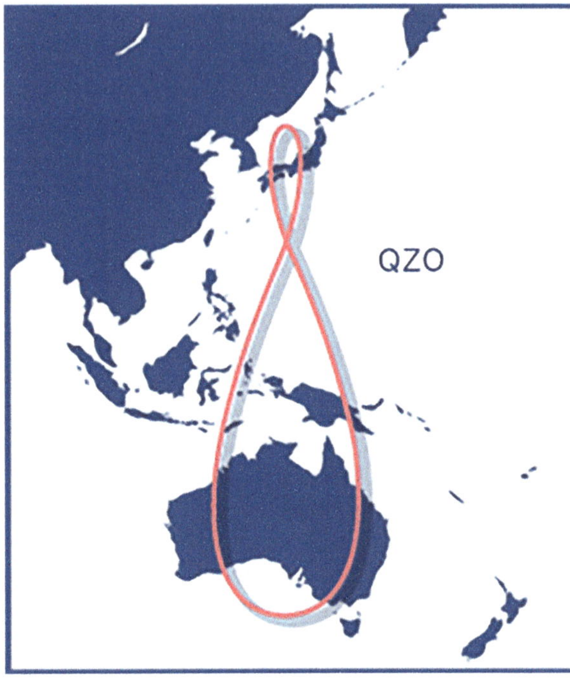

Fig. 2.1 Main area where QZSS "MICHIBIKI" is available. [Source: Cabinet Office, Government of Japan]

2.4 GSI Maps: Digitization of Topographic Maps and Web-Based Applications

Base maps in Japan are still being updated by the GSI. The most important 1:25,000 topographic maps were produced based on paper maps for more than 4400 areas of the country until the early 2000s. In addition, several thematic maps, such as land use, land conditions, volcanic land conditions, and maps of active faults in urban areas, have been produced, and these maps have become effective tools for disaster prevention activities and national development. These maps are also distributed in paper and digital media. Web maps called "GSI maps" are now widely used. Since 2009, topographic maps have been maintained on a digital map database which seamlessly links the entire country, together with orthoimages and place name information maintained by the GSI. These are positioned in a database called the base registry, which is the foundation of society.

GSI Maps (Fig. 2.2), which have been developed and operated by the GSI since 2003, became the current tile-based web map service in 2013. The main feature of GSI maps is that more than 100 base maps, aerial photographs, and thematic maps, such as land use, can be easily viewed with a web browser. The major advantages of GSI maps are that they can be shared widely based on the open data policy, and the GSI maps themselves are built using open-source software (Fujimura 2015).

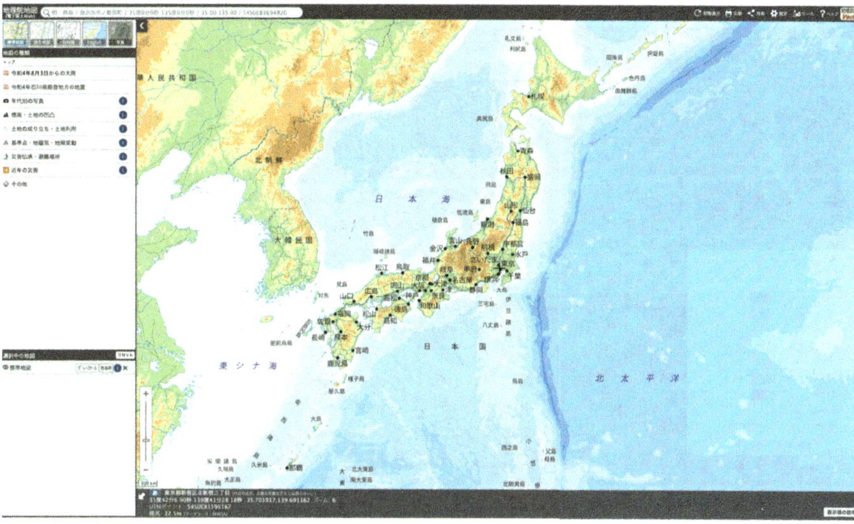

Fig. 2.2 GSI maps website. [Source: https://maps.gsi.go.jp/]

There are many map contents provided, and a map of the entire country is seamlessly maintained at a scale equivalent to 1:25,000. Urban areas and some map contents are at a scale equivalent to 1:2500 to 1:5000 to enable users to grasp the detailed land-use situation. According to Kamimura (2019), in addition to aerial photographs of the past and present, thematic maps including disaster information such as active fault maps are available, as well as user-customized color-coded elevation maps based on digital elevation models and 3D views, which can be downloaded in standard triangle language (STL) or virtual reality modeling language (VRML) format for output on a 3D printer (Figs. 2.3 and 2.4).

Second, GSI Maps is provided as open data by slippy map tilenames (XYZ), making it easy to incorporate into various web map services. In 2019, in addition to the tile service in image format, a trial provision of vector tiles started (Kamimura 2019). Through these elemental technologies, the recent creation of the United Nations Vector Tile Toolkit (UNVT) has also led to the development of a toolkit that can be used offline and on the Raspberry Pi, which is expected to expand the possibilities for further data utilization.

Third, the construction and operation of GSI maps are mainly based on open-source software, and it is groundbreaking in that it was the first Japanese government agency to provide source code on Github (https://github.com/gsi-cyberjapan/gsimaps). This is a breakthrough because it allows for the sharing of GSI maps technology and open collaborative communication with private sector engineers to improve the program (Mogi 2019). As of the end of January 2022, the project page on Github has been forked (a copy of the entire source code) more than 600 times, and feedback from about 100 contributors has been useful for improving and upgrading the functions of GSI maps.

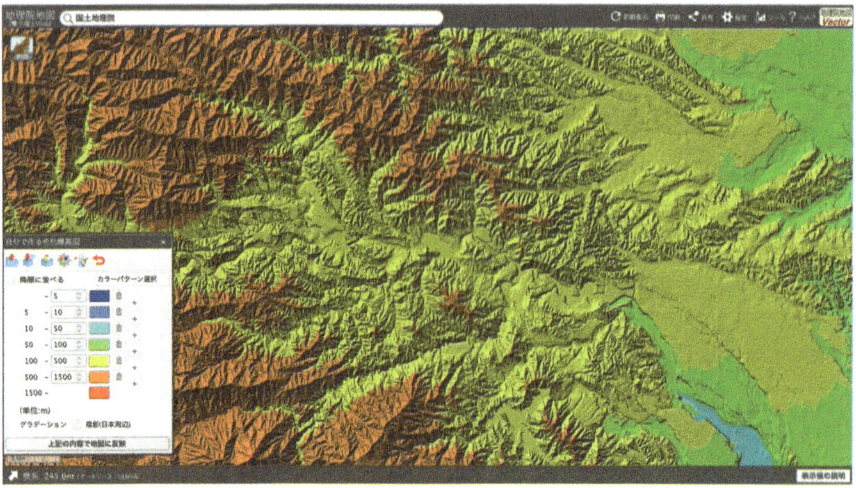

Fig. 2.3 "Your own make color-coded elevation map" in GSI maps. [Source: https://maps.gsi.go.jp/]

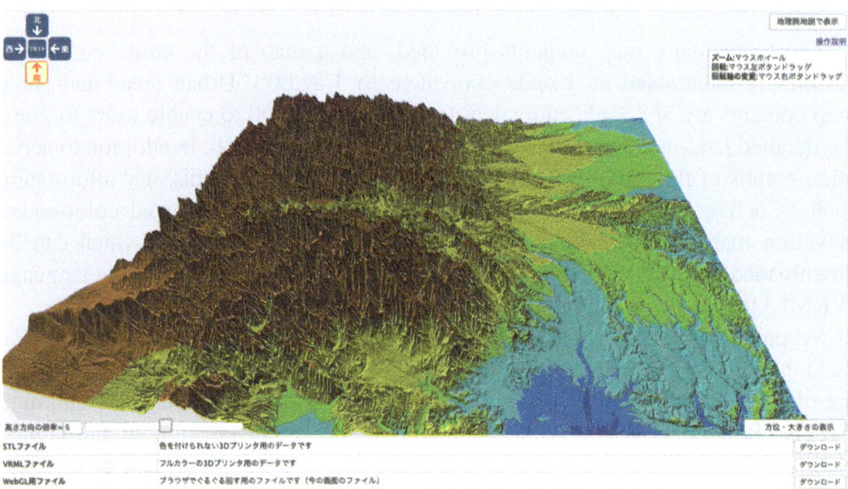

Fig. 2.4 3D topographic map visualization and export function. [Source: https://maps.gsi.go.jp/]

2.5 Web-Based Map Services and Utilization of Old Japanese Maps

In recent years, in addition to the provision of modern base maps such as topographic maps, it has become common practice to interactively compare changes in local land use on maps. Moreover, the preservation and utilization of valuable materials with cultural heritage value through digital archiving has attracted attention. Old Japanese

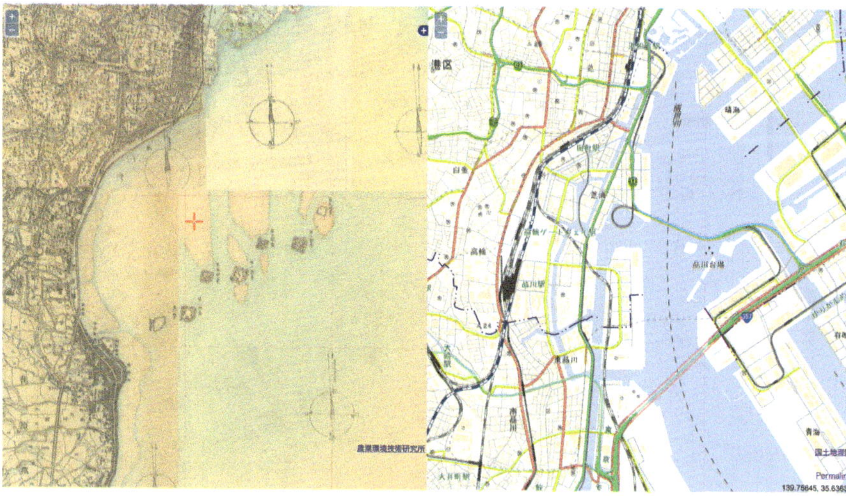

Fig. 2.5 Historical Agro-Environmental Browsing System. [Source: https://habs.rad.naro.go.jp/]

maps, which were produced around the Edo period (1603–1868), are also an important resource and are archived on the Internet by various organizations.

Figure 2.5 shows the Historical Agro-Environmental Browsing System of the Institute for Agro-Environmental Sciences, National Agriculture and Food Research Organization (NARO), which is a digitized topographic map of the Kanto region from the early to mid-Meiji era. The left side of the map shows the Odaiba area of Tokyo in the 1880s, and the small islands in the center of the map represent a gun battery. These places were built to strengthen the sea defenses of the Edo Shogunate in response to the arrival of Perry in 1853; however, land reclamation had not progressed during the Meiji period, and traces of the emplacements were left behind. In addition, to quantitatively analyze the land-use change from the 1880s to mid-1970s in the Kanto region, Iwasaki et al. (2015) constructed a grid-based database, which allowed them to estimate the area of wastelands such as grasslands and shrubs. Thus, historical maps are not only of documentary value but also tools that can quantitatively reveal spatial transitions.

Since the maps on this website are provided as open image tile data, they have been adopted by various websites of different organizations and have become one of the standards for historical maps. In addition to the above, the "Old Map Collection" of the GSI, archives of research institutes, regional museums, libraries, etc. are actively promoting the release of their data through the web. A large-scale archive site that integrates data from public institutions is "Japan Search," which was officially opened to the public in August 2020. It is operated by the National Diet Library and allows cross-searching and browsing of the archives of collaborating institutions. For example, as shown in Fig. 2.6, images and titles of valuable old maps from the Edo period are described, and links to map contents that can be viewed on the website are also provided.

Fig. 2.6 *Edo kirie-zu area maps* in MapWarper. [Source: https://mapwarper.h-gis.jp/layers/25]

Unfortunately, many of these maps are not aligned with the current maps, so it is not easy to superimpose them on a web map or geographical information systems (GIS), as shown in Fig. 2.5. As one of the means to overcome this problem, the "Japan Map Warper" (https://mapwarper.h-gis.jp/), which is operated by the Department of Geography, Ritsumeikan University, was set up. It is an open source-based sharing platform for old maps of Japan (Fig. 2.6). It allows users to upload map images on the website, align them with current map images, and share them as GeoTIFF, KML, or tile data. Figure 2.6 shows a mosaic of approximately 20 Edo kirie-zu (cutting and painting) area maps produced in the 1850s, which were mosaicked and combined into a single sheet. In a similar effort, there is an open-source map viewer library called "Maplat" (https://github.com/code4history/Maplat/wiki), which can be used to combine old maps with current location coordinates without distorting them and has been adopted by several smartphone applications for sightseeing maps. It has been used in several smartphone applications for tourist maps.

Thus, digitization of analog maps, including old maps, and sharing services for data that do not have accurate positional coordinates, to be able to use them in GIS, are gradually beginning to take place.

2.6 Conclusions

This chapter described the historical background of surveying and mapping in Japan's national institutions and explained the basic methods used in mapmaking. Further, the open provision of base maps is also promoted in Japan, mainly by the GSI, and some of its functions are mentioned using GSI maps as a case study. In Japan, many old maps have been digitally archived as a clue to understanding past land use, and it is becoming popular to overlay this information on current maps. It also became apparent that many of these websites are based on open-source technologies, which should be acknowledged as a distinctive feature.

References

Fujimura H (2015) An effective synchronization method of GSI tiles using a GSI tile list. Bull Geospat Inf Authority Japan 63:1–5

Iwasaki N, Spraguea D, Fujita N, Yamaguchi H (2015) Developing a land use database of the Kanto Region, Japan in the 1880's. In: Free and open source software for geospatial (FOSS4G) conference. Proceedings, 15, article 41. https://doi.org/10.7275/R5FT8J83

Kamimura K (2019) GSI maps: showcase of national geospatial data. In: Abstracts of the ICA, vol 1, p 1. https://doi.org/10.5194/ica-abs-1-158-2019

Mogi H (2019) The advancement of utilizing geospatial information via GSI maps. In: Abstracts of the ICA, vol 1, p 1. https://doi.org/10.5194/ica-abs-1-252-2019

Rooney A (2015) The story of maps: putting the world in perspective. Arcturus, London

Une H (2021) Current form of map making: measure the earth and draw a figure. Kodansha. (in Japanese)

Chapter 3
Various Representations and Processing Techniques of Geospatial Information: Appropriate Control of Big Data Behind

Yoshihide Sekimoto

Abstract In this chapter, we will learn about the basic representations and types of processing of various types of geospatial information. We will also learn how the processing is becoming more sophisticated due to recent needs to handle not only static data but also dynamic data including web-based and temporal information.

Keywords Shapes · Attributes · Superimposition · Geocoding · Search · Metadata · Map tiles · Spatio-temporal information · Standardization

3.1 Basic Representation of Geospatial Information

3.1.1 Graphics and Attributes

The term "geospatial information" is used to describe information about real-world events. Geospatial information, which is information about geographical objects, is basically represented by a combination of shapes and attributes. Various geographic objects, i.e., roads, buildings, rivers, trees, people, and cars in the real world, have complex shapes, but most of them can be represented with shapes based on points, lines, and surfaces, and various attribute information associated with them. Figure 3.1 shows various facilities in a town, such as a department store as a surface of a building, an optical cable buried underground as a line, and a regulatory sign as a point. This way, for the first time, you can click on the map you see and see the information you need.

This basic representation has not changed since the 1970s when GIS began, and this is stored in a file according to the data format. While the Shapefile file format adopted by ESRI's ArcView2.0 in 1996 is still in use, various file formats have been

Y. Sekimoto (✉)
Center for Spatial Information Science, University of Tokyo, Tokyo, Japan
e-mail: sekimoto@iis.u-tokyo.ac.jp

Y. Sekimoto, Y. Kawahara (eds.), *Utilization of Geospatial Information in Daily Life*, New Frontiers in Regional Science: Asian Perspectives 65,
https://doi.org/10.1007/978-981-19-6905-8_3

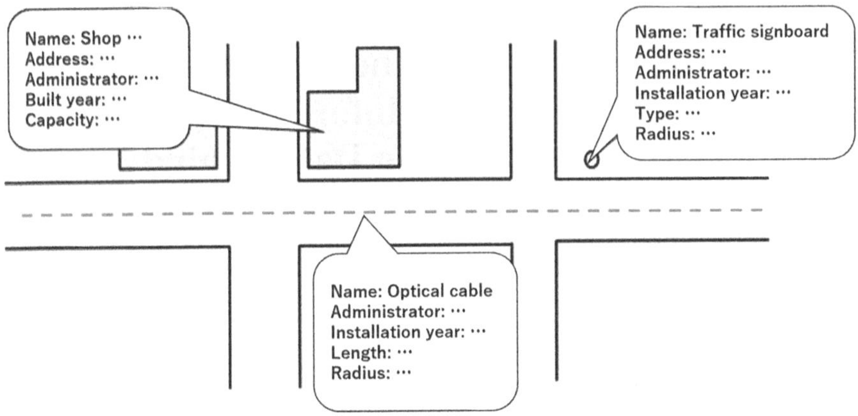

Fig. 3.1 Representation of geospatial information using graphics and attributes

devised and internationally standardized, such as Google's KML file for the Internet age since the 2000s (see Sect. 3.3 for standardization).

The above formats are called vector formats because they can be scaled up or down according to scale. On the other hand, things such as camera and satellite images can also be called geospatial information. These are called raster formats, in which attribute information such as landscape, ground surface condition, plant condition, population density, etc. is linked to a mesh represented in a grid pattern.

Also, the same geographical feature is not always represented by the same figure. Even the shape of the department store I mentioned earlier is displayed in different ways depending on the scale of the earth, Japan, the city, or your own surroundings. This is called LOD (level of detail).

3.1.2 Overlaying by Layers

One of the interesting and easy-to-understand aspects of geospatial information is that it can be overlaid with various types of information using location information as the key. Even information with completely different characteristics can be overlaid to reveal new things. For example, Fig. 3.2 shows a service provided by Tokushima Prefecture on its website called the Comprehensive Map Provision System. By simply overlaying a background map showing rivers, railroads, roads, etc., with an elementary school construction zone and landslide danger areas, we can see areas that require some attention. While there is no need to be more sensitive than necessary, it is important to be able to make calm judgments based on such objective overlays of information.

In addition, this kind of overlapping of layers is not limited to a single system, but as the Internet age progresses, it is also possible to overlap different systems on the Web, such as WMS (Web Map Service) and WFS (Web Feature Service). WMS and

Fig. 3.2 Overlay of background map, elementary school districts, and landslide hazard areas. (Cited from https://maps.pref.tokushima.lg.jp/)

WFS were proposed by the OGC (Open Geospatial Consortium), an industry group promoting international standardization, in 1999. The former refers to the layer level, while the latter refers to the level of individual geographic features, which enables superimposition without the need for advanced Web programming. The Ministry of Land, Infrastructure, Transport and Tourism and the National Research Institute for Earth Science and Disaster Prevention, for example, are actually implementing such a service in Japan.

3.1.3 Software to Operate Geospatial Information

There are now many software and web systems for manipulating files that store geospatial information, and ArcView (now ArcGIS), mentioned in Sect. 3.1.1, is the first commercial desktop application that is widely used. On the other hand, free software is also increasing due to the spread of open-source software in recent years. QGIS, which was just released in 2009, is rapidly gaining popularity because it has almost the same functions as commercial software (Fig. 3.3).

There are also many server software that realize WebGIS from commercial software, including the server version of ArcGIS, to free software such as MapServer and GeoServer, which are often combined with JavaScript open-source software such as OpenLayer to display maps in a browser. It can be said that there is a wide range of options for different uses.

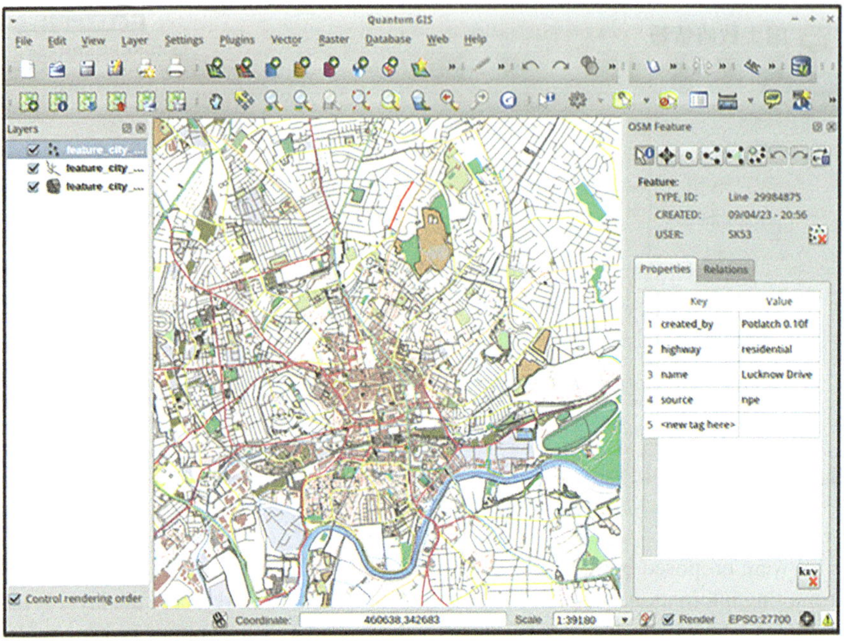

Fig. 3.3 QGIS operation screen. (Cited from QGIS tutorial page: http://live.osgeo.org/ja/quickstart/qgis_quickstart.html)

3.2 Processing Technology for Geospatial Information

3.2.1 Processing Technology

As described in the previous section, geospatial information created by various entities is processed into various forms by users. The method of doing so is case-by-case, but recently, conversion libraries have been increasing, making it easier to do so. Rather than file format conversion, I would like to introduce geocoding to obtain location information by latitude and longitude, which is essential for geospatial information.

As mentioned above, location information is essential for geospatial information, but not all information is expressed in latitude and longitude, which are easily interpreted by computers. However, not all geospatial information is expressed in terms of latitude and longitude, which are easy for computers to interpret. Geospatial information is often recorded and expressed in a form that is easy for humans to understand, such as in surveys and questionnaires, and may be expressed in terms of familiar names such as addresses, facility names, or street names. In such cases, it is necessary to convert the data into latitude and longitude in order to make it easier for computers to interpret, which is called geocoding.

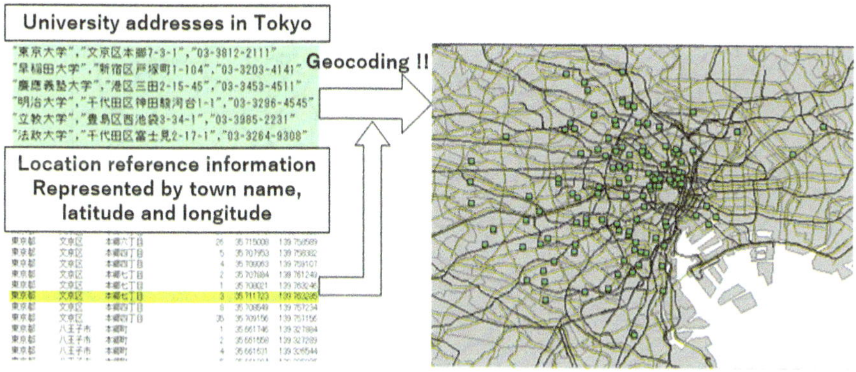

Fig. 3.4 Geocoded, latitude-longitude, and mapped place name information from the university list. (Created based on http://newspat.csis.u-tokyo.ac.jp/geocode/)

Figure 3.4 was created based on information from the address-matching service (http://newspat.csis.u-tokyo.ac.jp/geocode/) at the University of Tokyo's Research Center for Spatial Information Science, but it is possible to map a list of universities in the 23 wards of Tokyo with their names, addresses, and phone numbers. Is it possible to map the list of universities in Tokyo's 23 wards with their names, addresses, and phone numbers on a map? In this case, a place name dictionary plays an important role. The word "dictionary" may sound difficult, but simply put, it is a table of correspondences between place names and latitude and longitude. The dictionary of place names used here is called "Location Reference Information" published by the Ministry of Land, Infrastructure, Transport and Tourism (MLIT), and it literally records the expressions at the district level, or more detailed district level, and the latitude and longitude of the representative points.

3.2.2 Search Technology

Once created, geospatial information is stored in a file or database and searched in various usage scenarios. For example, it can be used to display the relevant area when scrolling on a map screen, or to find the shortest route to a destination. For example, when you scroll on a map screen and want to know the shortest route to your destination. This technology has become commonplace with the advent of Google Maps, etc., but its origins can be traced back to computational geometry started by Shamos et al. in the 1970s. This technique is used to find the range of influence of a point, to list up the shapes included in a certain area, and to calculate the cost of a network of connections to find the shortest path. It literally clarified the calculation method and efficiency of handling shapes on a computer. Computational geometry is systematically summarized in an easy-to-understand manner in, for example, Iri and Koshizuka (1986).

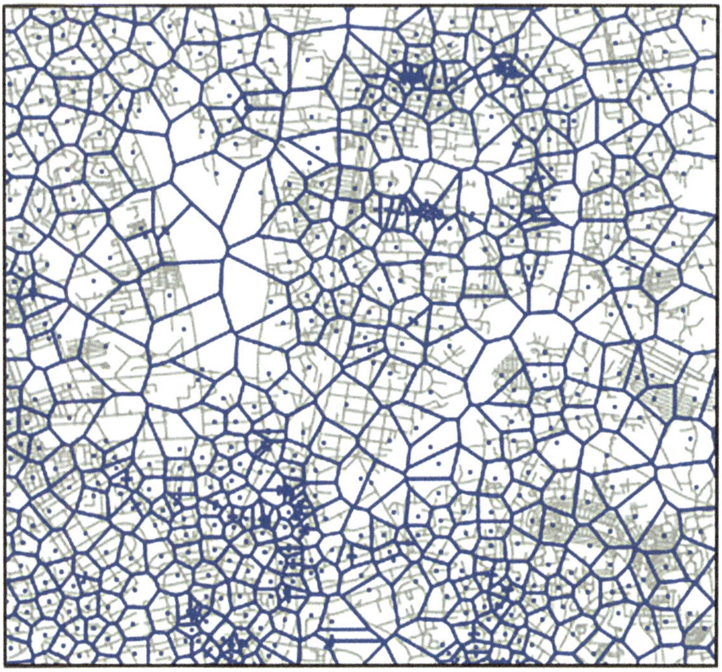

Fig. 3.5 Voronoi diagram depicting the coverage of mobile base stations in Dhaka city (blue dots: mobile base station locations, blue line: Voronoi diagram, gray line: road network. Created based on Hasegawa et al. 2014)

For example, Fig. 3.5 shows a Voronoi diagram of the sphere of influence of each cell phone base station in Dhaka city, with the road network superimposed in the background. The Voronoi diagram creates an area where each point, in this case the base station point, is closest to an arbitrary point, and can be created by connecting the perpendicular bisectors to each neighboring point. In this way, it can be clearly seen that the areas with high density of road network and high density of people also have high density of mobile base stations, and therefore, the area of influence is narrow.

3.2.3 Metadata Creation Technologies for the Web

As computer performance increases, data increases, and the Web becomes more advanced, the need for more real-time search results increases, and it becomes more important to prepare searchable information in advance.

This was around the year 2000, when Google's search technology, known as Web 2.0, spread throughout the world. It is said that Google handles billions of pages, and the indexes are 100 petabytes in size, and the time spent to build them is over

1 million hours. This is a considerable amount of time. Of course, there was indexing technology for databases before that, but the main focus was on thorough practical technology for handling data from all over the world (see (Google Search Education n.d.) for details).

The same thing is being done for geospatial information. Each geospatial information such as roads and buildings can be assigned a code such as a polygon of an administrative district or a mesh of a region in advance, or a network connection relationship in advance, which makes it easier to search accordingly.

3.2.4 Map Tiling Technology for the Web

Although this is a different concept from metadata in search, as it has become commonplace to browse maps on the Web, it has become important to prepare appropriate maps according to scale and display them at high speed, although the overall map information is detailed and enormous. A single tile is a fixed size of 256 pixels square, and as shown in Fig. 3.6, zoom. Each tile is fixed at 256 pixels square, and as shown in Fig. 3.6, zoom level 0 represents the entire map of the earth with one tile, zoom level 1 with four tiles, and zoom level 2 with 16 tiles. Since a specific zoom level is always specified when displaying on the Web, it is possible to handle this by simply reading one or several tile images depending on the center coordinates. This technology was also developed by Google in conjunction with GoogleMaps in 2005, but is now an industry standard technology used by many map services.

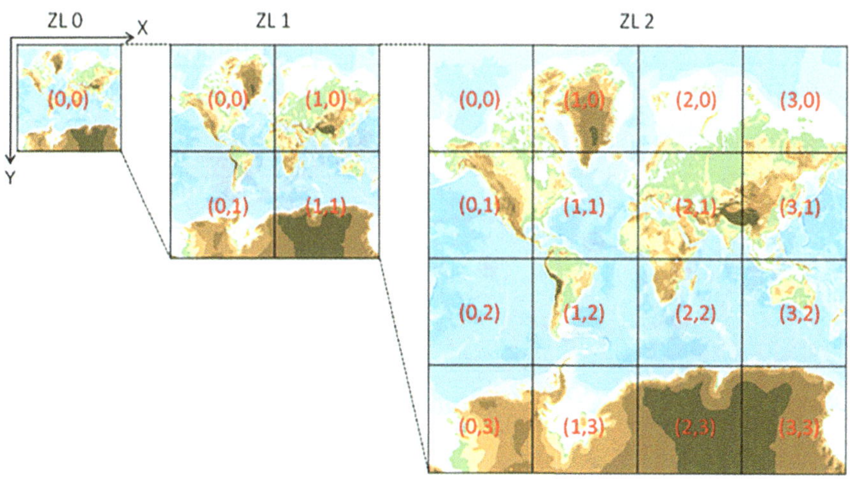

Fig. 3.6 Map zoom level and tile coordinates. (Cited from https://maps.gsi.go.jp/development/siyou.html)

3.3 Representation of Moving Objects Including Time and International Standardization

3.3.1 What Is Spatio-Temporal Information?

We have been talking about spatial information so far, but how do we represent geographic objects that change with time, such as people and cars? The answer is simple: put a time on it. However, as we have always assumed that spatial information is always in the same place, the amount of data increases with the change in time. For example, if a person's position is recorded every minute, the amount of data would be 1440 times that of a day without movement. Also, if you record the location every second, the number increases by 86,400 times. The GPS chips built into smartphones are often said to be 100 or 200 Hz, so if you record everything, the number increases another 100 or 200 times. This is the reason why it is called "big data" in recent years.

Figure 3.7 shows the position coordinates actually measured by the positioning function in a smartphone. The coordinates are recorded while moving outdoors and staying inside a building, with some errors.

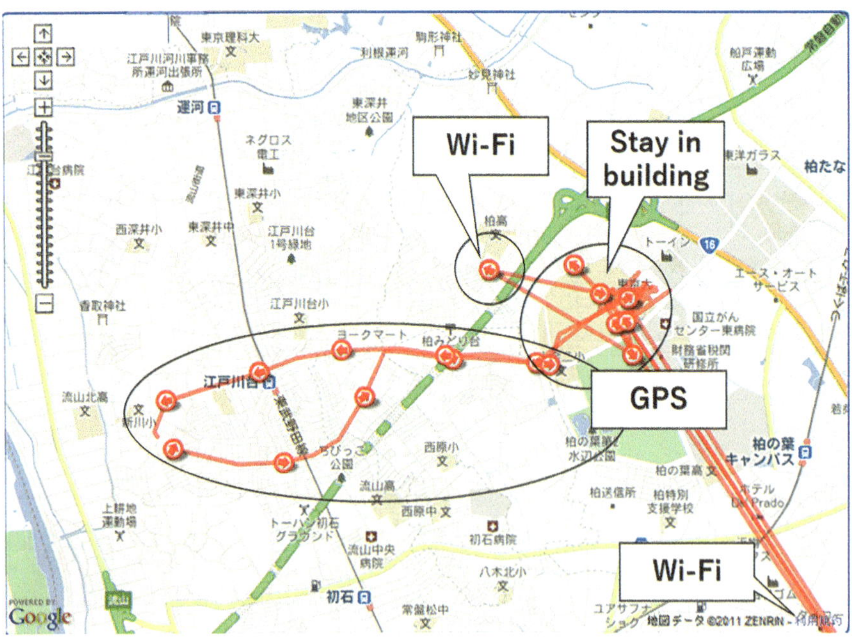

Fig. 3.7 Mobility and stay status of a person. (Created based on Sekimoto et al. 2011)

3.3.2 Large-Scale Data Processing Technology

In the case of huge data such as spatio-temporal information, it is even more important to have metadata creation technology for search, as described in Sect. 3.2.3 (Fig. 3.8). For example, if you want to know the movements of people in Tokyo on March 11, 2011, if you do not do anything, you will need to search the GPS data of all the people on that day. In the case of (Witayangkurn 2013), about 1.5 million people were recorded in a CSV file containing 9.2 billion lines of GPS data for 1 year (600 GB in total). This is partly due to the use of a distributed processing system called Hadoop (n.d.), but it is also reported to be 150 times faster than processing with spatial indexing in a Regular Relational Database (RDB).

3.3.3 Standardization of Geospatial Information

Finally, I would like to touch on the standardization of geospatial information that has been discussed in this chapter. As I mentioned in the beginning, standardization is a way to combine formats if everyone is thinking alike. With regard to geospatial information, TC211 of the International Organization for Standardization (ISO) started to develop de jure standards in 1994, and the Open Geospatial Consortium (OGC), a nonprofit organization in the United States, started to develop de facto standards at the same time. Initially, the main topic was static geospatial representation, but with the spread of the Web, the topic has shifted to the standardization of things like WMS and even moving objects.

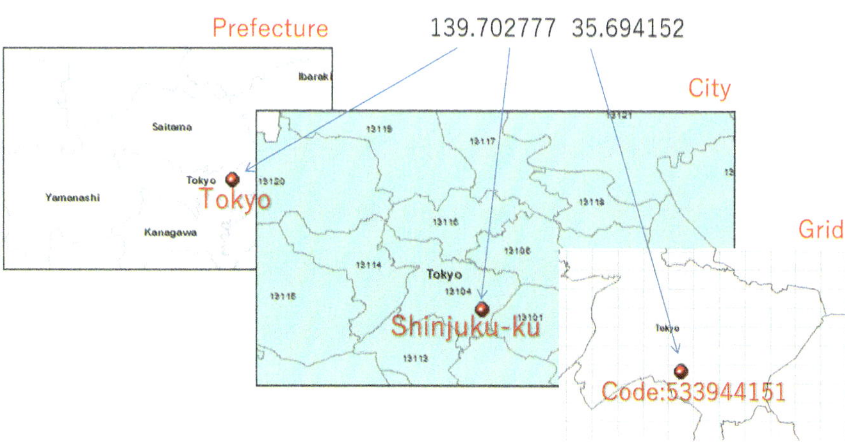

Fig. 3.8 Indexing of administrative codes and mesh codes into GPS data

3.4 Conclusion

In this chapter, we learned about the basic representations of various types of geospatial information and the types of processing, and we learned how the processing is becoming more sophisticated as the needs of the times change and the scope of geospatial information becomes not only static but also dynamic, including temporal information.

References

Google Search Education (n.d.) Google's homepage. https://www.google.com/insidesearch/searcheducation/

Hadoop (n.d.). http://hadoop.apache.org/

Hasegawa Y, Sekimoto Y, Kashiyama T, Kanasugi H (2014) Transportation Melting Pot Dhaka: road-link based traffic volume estimation from Sparse CDR Data, The 1st International Conference on IoT in Urban Space (Urb-IoT 2014), Rome

Iri M, Koshizuka T (1986) Computational geometry and geospatial information processing. Kyoritsu Shuppan, Bunkyo-ku

Sekimoto Y, Horanont T, Shibasaki R (2011) Commentary: current trends in people flow analysis technology using mobile phones. Inf Proc 52(12):1522–1530

Witayangkurn A (2013) A study on human activity analysis with large scale GPS data of mobile phone using cloud computing platform. Doctoral thesis, University of Tokyo

Part II
Mobile Communications and Positioning

Chapter 4
Ground-Based and Space-Based Positioning

Yasuhiro Kawahara

Abstract Positioning technology is essential to determine a location on the Earth. Real-time display of a user's or a driver's location on smartphones or car navigation systems is made possible every day by the application of multiple positioning technologies. This chapter describes the positioning methods and their applications, focusing on radio-based positioning methods and their application in indoor positioning, which are still being developed. GNSS, which has become a synonym for positioning in daily life, has also been explained, including its underlying mechanism and pattern of utilization.

Keywords Wireless positioning · GNSS · Indoor location measurement · xDR · SLAM

4.1 Satellite-Based Positioning

4.1.1 History of Radio Navigation

When using the navigation function of a car navigation system or smartphone, navigation systems need to know their own position. This section explains how a mobile terminal in a location information system uses radio waves to determine its own location.

In ancient times, the compass, a tool for determining direction, and the sextant, a tool for measuring latitude, were used with geographic navigation using landmarks, astronomical navigation using the motion of celestial bodies, and other techniques to determine one's location. In the eighteenth century, the development of the chronometer, a precision watch, made it possible to measure the longitude and check one's position on a map.

Y. Kawahara (✉)
Division of Art and Sciences, The Open University of Japan, Chiba, Japan
e-mail: kawahara2@ouj.ac.jp

Y. Sekimoto, Y. Kawahara (eds.), *Utilization of Geospatial Information in Daily Life*, New Frontiers in Regional Science: Asian Perspectives 65,
https://doi.org/10.1007/978-981-19-6905-8_4

In the twentieth century, radio wave-based positioning methods were invented and applied for navigation. Long Range Navigation (LORAN), developed in the United States, was the first full-scale long-range radio navigation system to use hyperbolic navigation. Hyperbolic navigation is a positioning method in which the difference in reception time of the received radio waves from two radio transmitters is converted into a distance, a hyperbola is drawn, and the intersection of the hyperbola drawn using three points is used as the object's position. In positioning using signals from satellites, a method to determine the observer's position based on the Doppler shift of the measured signals was devised in the 1960s and applied to a navigation system called the Navy Navigation Satellite System (NNSS). This positioning method is a precursor to the global positioning system (GPS), which has become a synonym for location in modern society. Currently, radio wave positioning is used for indoor and land positioning as well as sea navigation. Satellite-based positioning is explained in the next section.

4.1.2 GNSS (Global Navigation Satellite System)

A global positioning system that uses satellites is called GNSS. This system uses radio waves transmitted by satellites to determine the latitude, longitude, and altitude of the receiver by analyzing the position of satellites, the time difference between the time of radio wave transmission and the time the radio wave arrives at the receiver, and the carrier wave. Several countries have launched satellites used for positioning systems (Table 4.1). Among these, GPS, GLONASS, Galileo, and BeiDou operate with approximately 30 satellites in global orbit and are globally compatible. In GPS, which has been in operation since the early days, four or more satellites managed by the U.S. Department of Defense orbit the Earth in six orbital planes, each at an altitude of approximately 20,000 km, with a cycle of approximately 12 h. It was opened for private use in 1993 and has become the standard GNSS in daily activities.

In GPS positioning, the distance from the receiver terminal to the satellite is the difference between the time of radio wave transmission of the GPS satellite and the reception time at the receiver terminal ($t_r - t_i$) multiplied by the speed of the radio wave (v). Thus, the position of the receiver terminal can be estimated by measuring the radio waves of the three satellites, as shown in the three equations in Fig. 4.2.

Table 4.1 GNSS in each country

Country	Name of GNSS
United States of America	GPS
Russia	GLONASS
EU	Galileo
China	BeiDou
India	NAVIC
Japan	QZSS (Quasi-Zenith Satellite System)

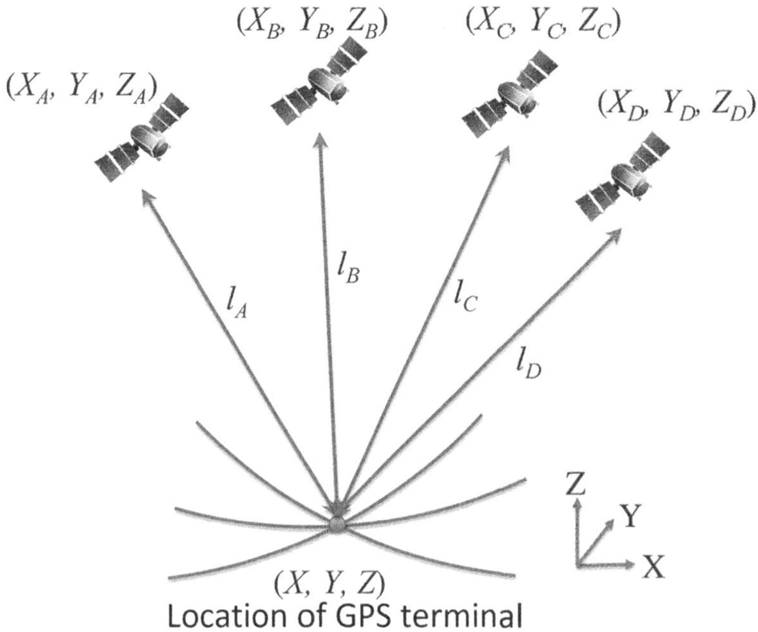

$$l_A = \sqrt{(X - X_A)^2 + (Y - Y_A)^2 + (Z - Z_A)^2} + ev = v(t_r - t_A)$$
$$l_B = \sqrt{(X - X_B)^2 + (Y - Y_B)^2 + (Z - Z_B)^2} + ev = v(t_r - t_B)$$
$$l_C = \sqrt{(X - X_C)^2 + (Y - Y_C)^2 + (Z - Z_C)^2} + ev = v(t_r - t_C)$$
$$l_D = \sqrt{(X - X_D)^2 + (Y - Y_D)^2 + (Z - Z_D)^2} + ev = v(t_r - t_D)$$

Fig. 4.1 Positioning by GPS

However, measuring this time difference requires an accuracy equivalent to that of an atomic clock for both the time of emission and the time of reception. It is necessary to find a solution using four equations, considering the error (e) at the time of reception by the receiving terminal, which is not as accurate as the GPS satellite time using an atomic clock (Fig. 4.1).

The information transmitted by GPS satellites includes time information, orbital information of the satellite (ephemeris), orbital information of all GPS satellites (almanac), and correction data, such as propagation delay due to the ionosphere. Each GPS satellite transmits 25 frames of this information. Each frame contains five subframes, and the information is transmitted in the order shown in Fig. 4.2. The TLM word is transmitted at the beginning of the subframe, followed by the HOW word. The TLM word contains the synchronization pattern and the HOW word contains the time information of the GPS signal.

The receiving terminal first receives an almanac from one satellite, determines the visible GPS based on the almanac, receives ephemeris from these satellites, and uses

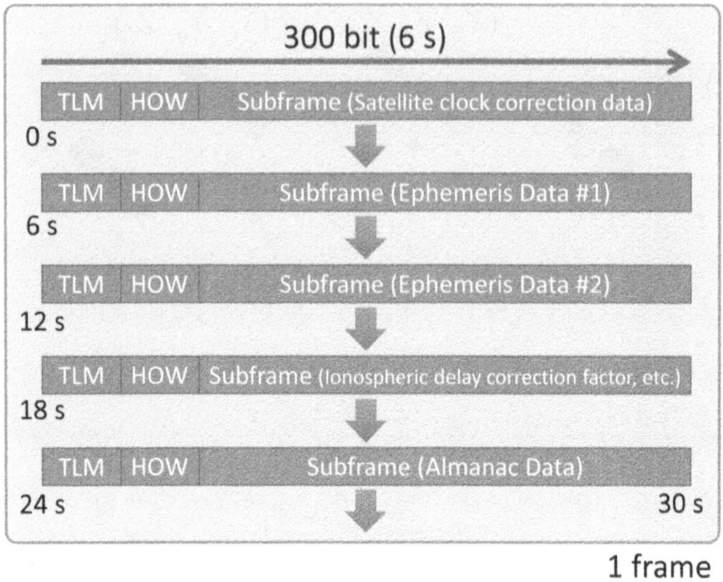

Fig. 4.2 Transmission information from GPS satellites

the satellite positions for determining its own position. A subframe consists of 300 bits, and each bit has a data length of 20 ms. The entire frame is 1500 bits, and the period of the frame is 30 s. Thus, the GPS receiver takes 12.5 min to receive the necessary data for these 25 frames when initially powered. If the receiving terminal has a valid almanac and ephemeris, it can start positioning immediately after powering on; this method is called hot start. This type of positioning system is called A-GPS (assisted GPS).

As GPS provides positioning in three dimensions, it also measures altitude. However, this altitude is not an elevation (orthometric height). The geodetic system used by the GPS for positioning is the WGS84 ellipsoid, and the altitude measured by the GPS is the altitude from this ellipsoid. It is thus called the height of the ellipsoid, and its relationship with the elevation is shown in Fig. 4.3a. The geoid height of the location must be subtracted to convert the GPS-measured altitude to elevation. For example, if this GPS altitude is used for surveying to establish a channel, when the altitudes of the two measured points have the relationship shown in Fig. 4.3b, the values of the elevations at the two points will be opposite to the relationship between the measured altitudes, and the gravity value at the measurement point will also be estimated inversely, resulting in water not flowing as planned. In addition, as GPS cannot provide positioning when there is no reception of radio waves from satellites, positioning in high-rise buildings or indoor environments is often difficult. Furthermore, when used for navigation while being carried, it is necessary to supply power to the GPS receiver module. Therefore, it is necessary to design a portable terminal that can withstand the purpose of navigation, considering the power consumption of the device.

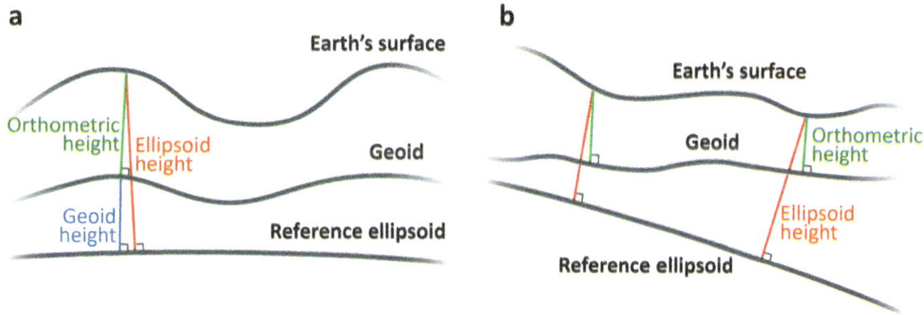

Fig. 4.3 (**a, b**) Relationship between ellipsoid height and elevation

4.1.3 Corrective Positioning

The following are examples of methods used to improve the accuracy of GNSS positioning.

Differential GPS (DGPS) positioning uses the difference (error) between the positioning results of a GPS receiver (reference station) at a location with a known position and the target position to correct the results of new GPS positions using another GPS receiver (mobile station). This enables positioning with an accuracy of approximately 1 m, unlike the GPS-alone positioning which has an accuracy of approximately 10 m.

In real-time kinematic GPS (RTK-GPS) positioning, the reference and mobile stations perform positioning simultaneously, and the relative position of the mobile station is calculated from the observed data. This method further improves the accuracy of the mobile station positioning.

In Japan, an electronic reference point is used as a form of reference station, which is installed to construct a high-precision survey network and monitor crustal movements. These electronic reference points are GNSS continuous monitoring devices installed at approximately 1300 locations throughout Japan by the Geospatial Information Authority of Japan (GSI), and they are used to monitor crustal movement caused by earthquakes by releasing observation data to the public in real time and by capturing changes in the observation data. Recently, private GNSS continuous monitoring stations have been established, and GSI has been evaluating their accuracy and registering them to promote their effective use. The development of these reference stations and utilization of observation data can improve positioning accuracy and facilitate various services, including smart agriculture.

4.1.4 Quasi-Zenith Satellite System

In Japan, the quasi-zenith satellite system (QZSS) is used as a solution for positioning in areas where connecting to GPS satellites is difficult. The QZSS consists of multiple satellites with orbits that pass near the zenith of Japan, enabling a single satellite to always be in the sky over Japan. With the satellite positioned almost directly overhead, the QZSS signal can expand the locations and times when positioning is possible, including mountainous areas and urban high-rise buildings, where the number of satellites required for positioning cannot be seen by GPS satellite signals. In other words, when only three GPS satellites are available, as shown in Fig. 4.4, it is possible to combine a QZSS satellite with another GPS satellite to improve positioning accuracy.

The orbit of a quasi-zenith satellite is an unobjective figure of 8, as shown in Fig. 4.5, and each satellite can stay directly above Japan for only 7–9 h. Therefore, it is possible to cover the sky over Japan 24 h/day by swapping three or more QZSS satellites at different times. As of 2021, following the first MICHIBIKI launched in 2010, three more were added in 2017, and four operations began in November 2018. A system of seven aircraft will be in operation by 2023 to acquire location information in urban and mountainous areas.

Sub-meter and centimeter-level navigation is expected to be realized by combining the positioning methods mentioned above. Seamless linkages with indoor positioning services and the ability to distinguish between sidewalks and roadways are also expected to contribute to automated vehicle driving technology.

4.2 Ground-Based Positioning

4.2.1 Types of Wireless Positioning

There are three main methods for estimating location using radio waves. Proximity uses the location of a single base station in the proximity of a mobile terminal for communication as the terminal location. Range based is based on the estimated

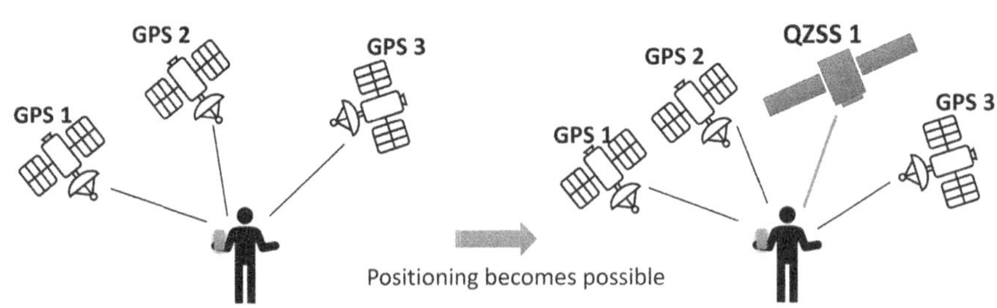

Fig. 4.4 Quasi-Zenith Satellite System

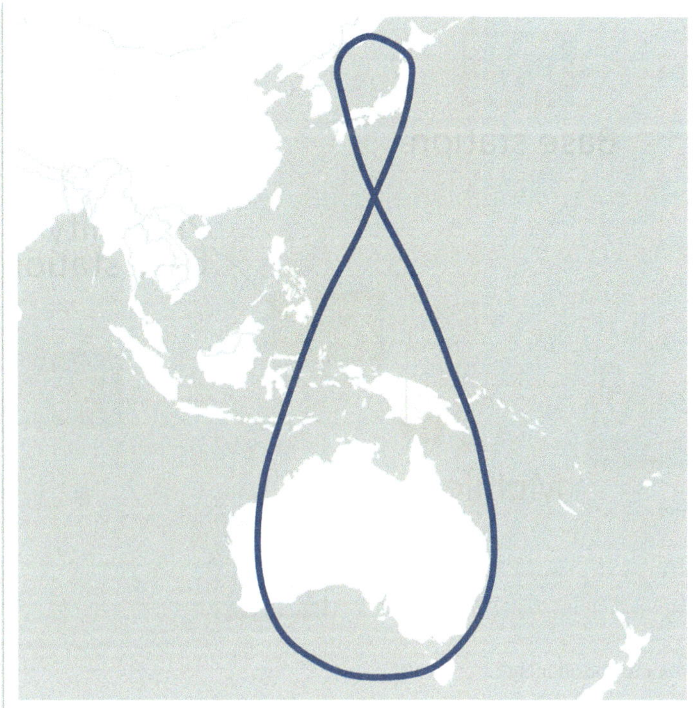

Fig. 4.5 Quasi-Zenith Satellite Orbits

distance range between the mobile terminal and the base station. AOA uses the reception direction of base station radio waves received by the mobile terminal.

Proximity is a method in which the location of the base station closest to the mobile terminal with which the mobile terminal communicates is used as the location of the mobile terminal. This is the simplest method. However, if the base station with which the mobile terminal is communicating is not the closest base station or if the base stations are located at large intervals, the positioning error will increase (Fig. 4.6).

Range-based positioning includes received signal strength (RSS), time of arrival (TOA), and time differential of arrival (TDOA). RSS estimates the position of a mobile terminal by estimating the distance from the mobile terminal to a base station using the property that the electric field strength of the radio wave transmitted by the base station, which is measured by the mobile terminal, decreases with the distance from the base station (Fig. 4.7). TOA is a method in which a base station measures the reception time of a signal from a mobile terminal, determines the location of the mobile terminal multiple times, and it is used for positioning with cell phones. TDOA is a method of estimating the position of a mobile terminal by measuring the difference in the arrival times of radio waves from the mobile terminal at three or more base stations. In TOA and TDOA, time measurement hardware needs to be

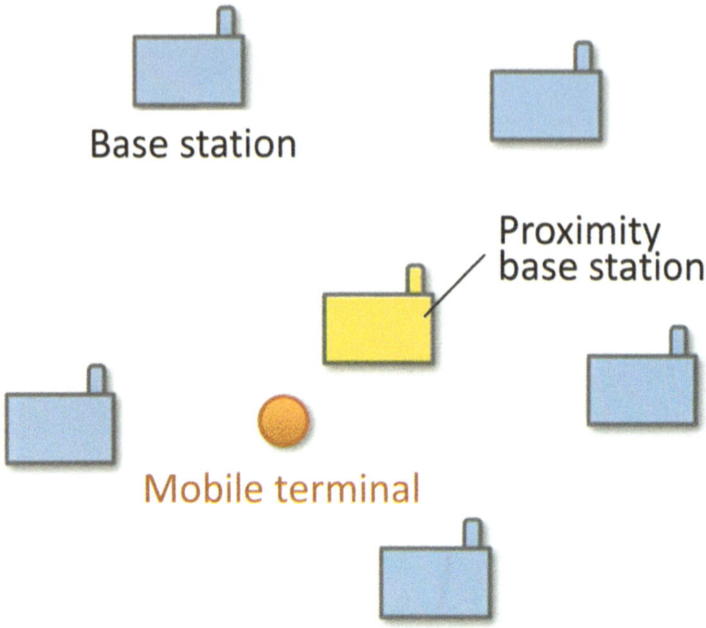

Fig. 4.6 Proximity positioning

Fig. 4.7 RSS positioning

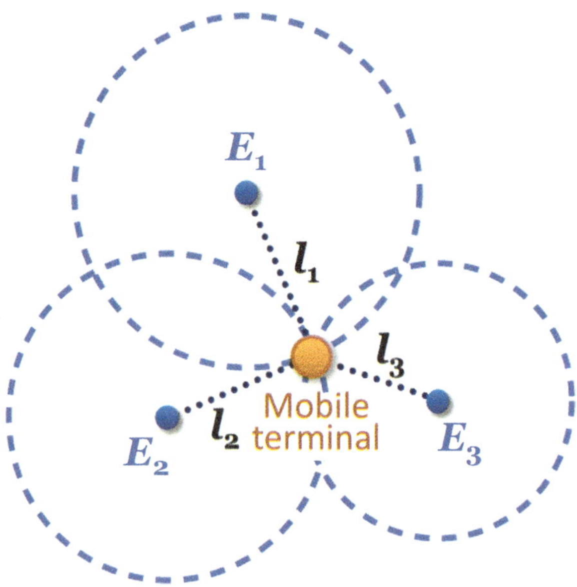

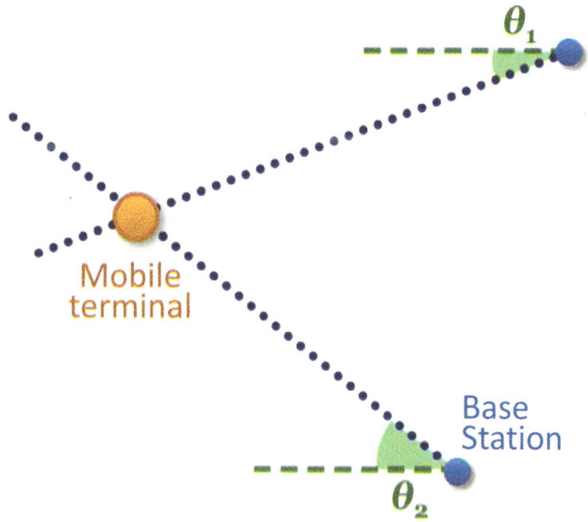

Fig. 4.8 AOA positioning

installed in base stations and mobile terminals to synchronize time among multiple detection devices and to measure time accurately.

AOA is a method of estimating position based on the direction of arrival of radio waves from a mobile terminal to a base station and the angles of beams from multiple base stations to the terminal, and it can be performed by a minimum of two stations. The base station antenna needs to be directional and is substantially affected by multipath. Therefore, its application in daily activities is limited (Fig. 4.8).

4.2.2 Wireless Positioning Applications

Positioning using received radio wave strength is referred to as the RSS method. It uses the property that the field strength of radio waves transmitted by a base station, which is measured by a mobile terminal, decreases with distance from the base station to estimate the distance from the mobile terminal to the base station, thereby estimating its position. This method uses the coordinates of a point whose location is already known and the distance from that point to the positioning point to obtain the coordinates of the positioning point. As shown in Fig. 4.9, the distance between a mobile terminal and a base station can be calculated from the strength of the radio wave as the location of the base station of a cell phone network is known in advance, and the farther the distance, the weaker the radio wave reaching the terminal. Although positioning is possible indoors and underground where cellular phone services are available, the accuracy depends on the spacing of the base stations, with an accuracy of approximately 300 m at 1 km base station intervals. This is due to errors in distance estimation caused by the reflection and diffraction of radio waves from buildings and terrain between base stations and mobile terminals. Several

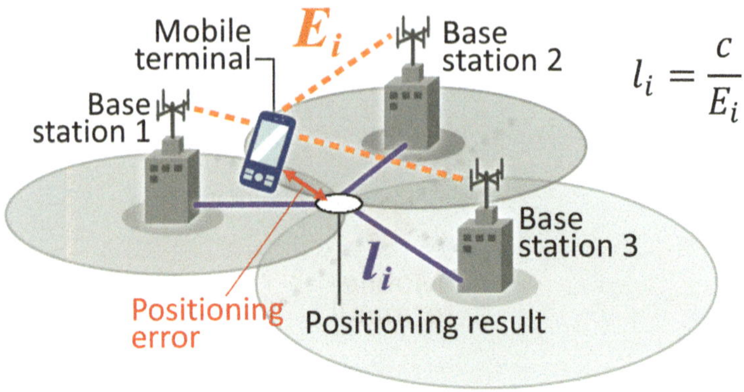

Fig. 4.9 RSS positioning using cellular network base stations

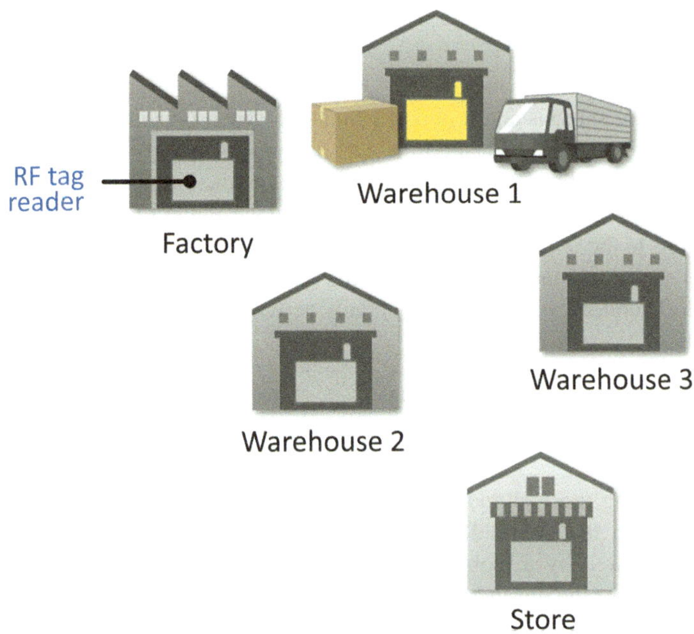

Fig. 4.10 Distribution management of goods using RF tags

methods have been used to reduce these errors. In addition, mobile communication terminals measure RSS values when communicating with base stations during ordinary data communications and calls. RSS positioning using cell phone base station radio waves can also be performed with low power consumption as no power is used for positioning.

Positioning with RF tags uses the proximity method. By installing RF tag readers at key locations, the position of the RF tag reader becomes that of the read object. As shown in Fig. 4.10, it is used to track goods and products and identify distribution

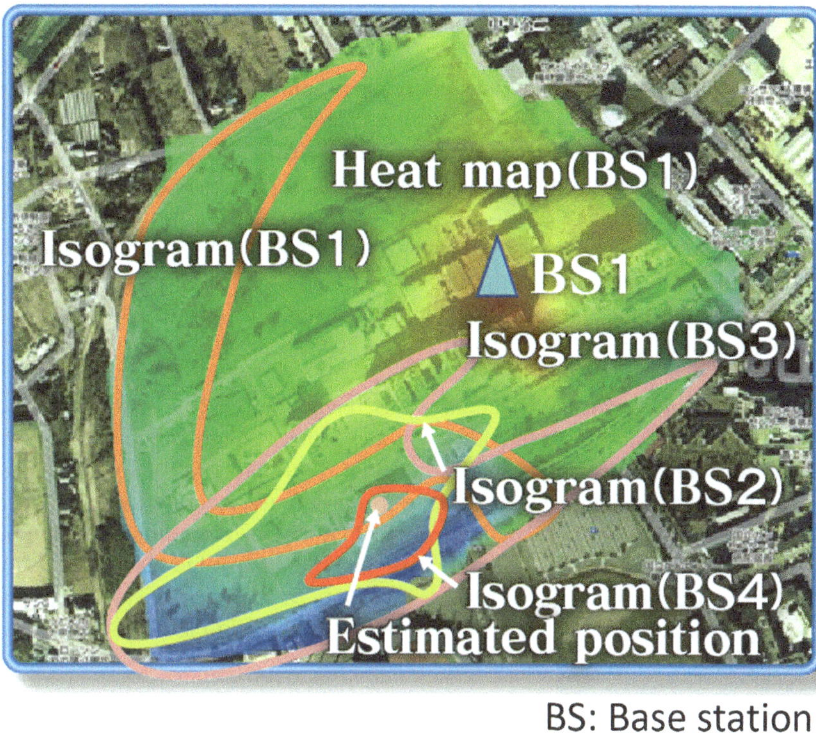

Fig. 4.11 Fingerprinting positioning (calculation by isopotential estimation)

routes and transportation conditions. It is beneficial for both the customer and vendor to know the current location of the package or goods, delay status, and future schedule. In combination with the GPS and cellular phone base stations, more detailed location information can be obtained. Fingerprinting is a positioning method that compares the radio wave strength from the base station, measured by the mobile terminal, with the radio wave strength distribution of the fingerprints available in the database.

Figure 4.11 illustrates the fingerprinting process. The heat map in the figure (the red area is strong, and the blue area is weak) depicts the propagation of the radio wave from one base station (BSi), which is measured in advance and stored in a database. The database contains a heat map of the radio propagation of each base station, which makes it possible to represent the radio strength of each base station measured at the mobile terminal as a contour line using the heat map of each base station (Yokoi et al. 2010). The positioning accuracy by fingerprinting can be improved by using the overlapping area of the contours as the location of the mobile terminal. This method is effective in reducing the effects of multipath and other positioning error factors on RSS positioning.

4.2.3 Dead Reckoning (xDR)

xDR is a positioning method that does not require radio waves. A familiar example in the present day is a car navigation system. When a car enters a tunnel during operation, the car navigation system loses satellite positioning. However, the car's self-position is indicated by the car navigation system even in a tunnel as it uses xDR, which locates the car's position relative to its own data gained from the car's travel. Travel data are acquired using onboard gyro sensors and speed pulse sensors. With the widespread use of smartphones, dead reckoning for pedestrians (PDR, pedestrian dead reckoning) has been developed using the movement status of people collected by smartphones equipped with various sensors such as accelerometers and gyro sensors.

The position indicated by the autonomous navigation is relative to the initial position. The self-position is estimated using the distance and direction information calculated by the sensor and the initial position. Therefore, except for the initial position, this method completes positioning using only the measured values of the onboard sensors, without using anything in the environment. A significant weakness of this method is that the errors in the distance and direction values estimated using the sensor values accumulate as the user moves. In many practical cases, the positioning accuracy is maintained in combination with radio-based absolute positioning methods.

4.2.4 Indoor Positioning

The practical application of indoor positioning, which is in demand at commercial facilities, hospitals, and warehouses, is currently being developed. For indoor wireless positioning, Wi-Fi access points or RF tags that can be attached to buildings or installations inside buildings are used as base stations. By managing the indoor locations of these base stations, many methods are used to estimate self-location using the ID and RSS values of each access point (base station) recorded by smartphones or other devices. In indoor positioning, there are many situations in which a combination of the above methods, called hybrid positioning, is required to improve positioning accuracy.

In the case of indoor positioning, indoor maps may not be available, and it is sometimes necessary to simultaneously measure both positioning and indoor geometry to give meaning to self-position in an indoor space. Therefore, a technology that simultaneously estimates self-location and maps the environment (SLAM) was used. In SLAM, the shape of the surrounding environment is measured using LiDAR (laser scanner) and cameras, and an environmental map is created. There are many situations in which SLAM can be effectively used, which are not limited to indoor positioning. A familiar example is the robotic vacuum cleaner for home use. By adding coordinates to an indoor map to manage indoor locations, it is possible to use indoor maps seamlessly with generally available outdoor digital maps. For example, by using an open street map for the outdoor map and a facility management map for the indoor map and by selecting multiple landmarks that exist in both maps and overlapping maps in GIS, the facility management map used for the indoor map can be placed in the geographic coordinate system. In Japan, indoor maps are increasingly available on digital maps on smartphone mapping platforms, such as Google Maps, Apple Maps, and Yahoo! MAP. Standardization of indoor map data is also being conducted; an example is Apple's indoor mapping data format (IMDF) (Nishio 2018).

References

Nishio N (2018) Indoor positioning and location information (Zukai Yokuwakaru Okunaisokui to Ichijoho). The Nikkan Kogyo Shimbun (in Japanese)

Yokoi N, Hosaka H, Kawahara Y, Sakata K (2010) High-accuracy positioning method using public wireless network based on recorded radio propagation characteristics. In: TENCON 2010-2010 IEEE region 10 conference, pp 2356–2360

Chapter 5
Observation from Space: Remote Sensing

Masahiko Nagai

Abstract This chapter introduces the basic concept of remote sensing, which is the observation of the earth from space using satellites, and its application fields. Also, the current status of remote sensing technology, which has been newly utilized due to recent technological developments, is explained.

Keywords Remote sensing · Earth observation satellite · Satellite constellation

5.1 What Is Remote Sensing?

5.1.1 Satellite Remote Sensing

Remote sensing means to observe and measure an object from a distance. Satellite remote sensing is a technology to observe the size, shape, number, and properties of objects and phenomena on the ground from space using sensors onboard earth observation satellites without directly touching the objects or phenomena.

There are several advantages to using satellite remote sensing technology. The first is the ability to safely observe from remote locations. For example, it is possible to observe disasters, conflicts, and other situations that are difficult to access from the ground from space. Also, deforestation in the Amazon can be observed using satellites without having to go directly to the site to study the causes of global warming. The second is the ability to observe large areas in a short period of time. For example, it is possible to observe global forest coverage and damages of huge disaster over country border. Third, it is possible to obtain information that cannot be captured by the human eye. Satellite remote sensing uses not only visible light, which can be seen by the human eye, but also infrared, thermal infrared, and

M. Nagai (✉)
Graduate School of Sciences and Technology for Innovation, Yamaguchi University, Ube, Yamaguchi, Japan
e-mail: nagaim@yamaguchi-u.ac.jp

Y. Sekimoto, Y. Kawahara (eds.), *Utilization of Geospatial Information in Daily Life*, New Frontiers in Regional Science: Asian Perspectives 65,
https://doi.org/10.1007/978-981-19-6905-8_5

Fig. 5.1 Satellite image of Tokyo observed by GRUS-1 satellite. (© AXELSPACE INC)

microwaves to observe the ground, making it possible to observe the temperature of the sea surface and, especially with microwaves, the ground at night and during rainy weather through the clouds. Fourth, the same area can be observed repeatedly. Satellite remote sensing has been accumulating data for about 50 years and can observe urban expansion, deforestation, and environmental changes over the entire world in a time series. Figure 5.1 is a satellite image of Tokyo on February 4, 2021, taken by the GRUS-1 satellite. Repeated observations will provide information on environmental changes and damage in the event of a disaster.

By analyzing satellite data, satellite remote sensing technology is being utilized in a wide range of fields, including disaster monitoring, resource management, forest coverage, environmental and weather information and environmental destruction, agricultural and fishery support, and infrastructure monitoring.

5.1.2 History of Remote Sensing

The history of satellite remote sensing is closely related to advances in aircraft technology and space exploration. The earliest history of satellite remote sensing dates back to the mid-nineteenth century, when aerial photographs were taken from balloons. In this era, there was no technology to transport observation sensors to space, so it started with taking camera pictures of a wide area from high altitude above in a balloon.

During World War I, aircraft technology also advanced, and many aerial photographs were taken for reconnaissance and other military purposes. In the latter half of

Fig. 5.2 ALOS-2 satellite. (© Japan Aerospace Exploration Agency (JAXA))

the twentieth century, the development of satellites made satellite remote sensing possible. Its civilian use began with the launch of the Landsat-1 satellite in 1972, which was launched by the United States. The data observed from Landsat-1 was made available to the public, and satellite remote sensing spread to the entire world. This data is still available today and provides information on the Earth's environment about 50 years ago.

Japanese satellite remote sensing began with MOS-1, launched in 1987. This was the first Earth observation satellite based on Japan's own technology for the purpose of ocean, agriculture, forestry, and environmental monitoring. This was followed by JERS-1 in 1992 and ADEOS in 1996. The Advanced Land Observing Satellite ALOS, which operated from January 2006 to April 2011, was used to assess the damage caused by the Great East Japan Earthquake and was a great success. Its successor, ALOS-2 (Fig. 5.2), launched in 2014, was used in various fields such as disaster monitoring using Synthetic Aperture Radar (SAR). The launch of ALOS-3 and ALOS-4 is also planned.

In recent years, satellite remote sensing has advanced dramatically, and private companies are now manufacturing, launching, and operating satellites. High-resolution images with a ground resolution of less than 1 m can now be acquired, and satellite imagery has become widely available to the public through services such as Google Maps. Furthermore, the development of nanosatellites weighing less than 100 kg is progressing, and Japan's nanosatellite development is at the top level in the world. Many emerging countries are working to develop and launch nanosatellites with Japan's technical support.

5.2 Earth Observation Satellite

5.2.1 Sensor Type

Sensors onboard earth observation satellites can be broadly classified into optical, thermal infrared, and microwave sensors. Figure 5.3 shows remote sensing sensors and their observation wavelengths. Optical sensors observe visible and infrared light. Optical sensors capture and observe visible light and near-infrared that are reflected when sunlight strikes an object on the ground. By examining the intensity of these reflections, it is possible to determine the distribution of plants, forests, and fields, as well as the condition of ground surfaces such as rivers, lakes, marshes, and urban areas. However, remote sensing using optical sensors cannot be used at night when there is no sunlight. In addition, when there are clouds, sunlight reflected from the ground surface is blocked by the clouds, making it impossible to observe the area under the clouds.

Thermal infrared sensors can capture and observe thermal infrared radiation emitted from the surface of the earth warmed by sunlight. They can also observe high-temperature areas such as volcanic activity and wildfires. By examining the intensity of the radiation, it is possible to measure ground surface temperature and sea surface temperature conditions such as heat islands. Remote sensing using thermal infrared sensors can observe the earth's surface even at night if there are no clouds.

Remote sensing using microwave sensors, called Synthetic Aperture Radar (SAR), can observe the roughness of the earth's surface by irradiating microwaves from a satellite to the surface and receiving the reflected waves from the satellite. Microwave remote sensing uses microwaves with longer wavelengths than visible

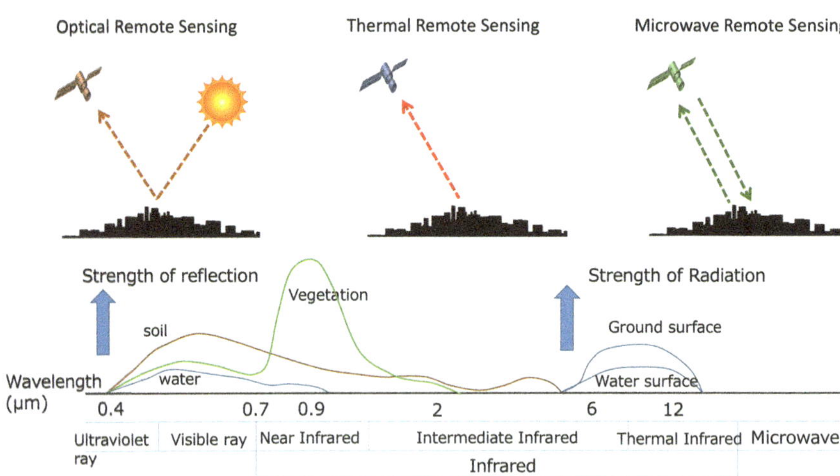

Fig. 5.3 Remote sensing sensors and observed wavelengths

light and infrared rays, so microwaves emitted from a satellite can penetrate clouds, allowing observations day and night, regardless of weather conditions.

5.2.2 Orbits of Earth Observation Satellites

Earth observation satellites used for remote sensing fly at an altitude of 500–600 km above the ground. Compared to the geostationary meteorological satellite Himawari, which flies in a geostationary orbit at an altitude of about 36,000 km, earth observation satellites fly at a very low altitude, but a low orbit is important to improve image resolution. In the case of remote sensing by optical sensors, it is important to have constant sunlight irradiating the observation site, so the satellite flies in a sun-synchronous orbit that enables observation at a constant local time (Fig. 5.4 (left)). In addition, in order to maintain a constant interval of days between observations of the same location, the satellite flies in a quasi-recurrent orbit in which the area to be observed shifts slightly each time the satellite travels around the earth and returns over the same location again a few days later (Fig. 5.4(right)). Most Earth observation satellites fly in a sun-synchronous quasi-recurrent orbit that combines these two orbits and are observed around noon.

Microwave remote sensing, on the other hand, does not require reflected solar radiation, so observations can be made day and night. However, since microwaves are emitted, the power requirements are large, and to ensure that sunlight is always available to generate electricity, most of them use sun-synchronous sub-recurrent orbits, which are observed in the early morning or evening.

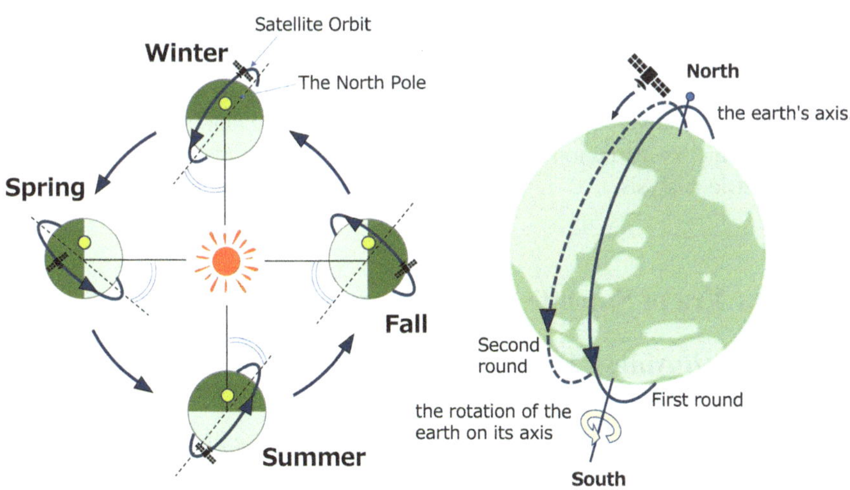

Fig. 5.4 Sun-synchronous orbit (left) and quasi-recurrent orbit (right). (© Japan Aerospace Exploration Agency (JAXA))

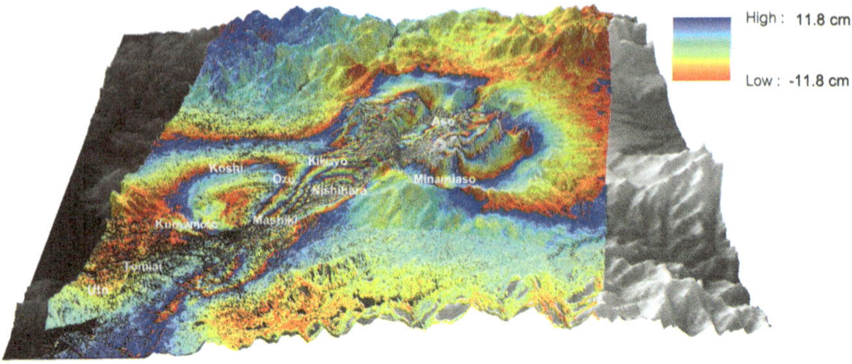

Fig. 5.5 Crustal deformation of the Kumamoto earthquake by SAR interferometric analysis of Daichi-2. (© Yamaguchi University)

5.2.3 Application Fields of Remote Sensing

The fields of application of satellite remote sensing are many and varied. It can be used to monitor disasters, examine abandoned farmland, which is increasing every year, monitor deforestation, and detect vessels navigating the oceans.

Satellite remote sensing provides a bird's-eye view of the Earth from space, and this is the greatest strength and role of satellite data. JAXA's ALOS-2 can observe the earth's surface with a sensor called Synthetic Aperture Radar (SAR), which can observe disasters at night and in bad weather.

There is a technique called SAR interferometric analysis that uses SAR to detect crustal deformation. In the 2016 Kumamoto earthquake, about 1 m of subsidence and 30 cm of uplift were detected on the north and south sides, respectively, of the Futagawa fault zone, which extends from the western slope of the Aso Outer Rim Mountains to the tip of the Udo Peninsula (Fig. 5.5). In addition, land subsidence due to excessive groundwater pumping in Bali, Indonesia, and land subsidence associated with tunnel construction in Chofu City, Tokyo, were also detected by SAR interferometric analysis.

5.3 New Era of Satellite Remote Sensing

5.3.1 New Business Through Remote Sensing Technology

In May 2017, the Cabinet Office of Japan released the Space Industry Vision 2030. The space industry is expected to be a driving force of the fourth industrial revolution and a new industrial frontier. The Japanese government aims to double the overall market size of the space industry by 2030, from 1.2 trillion yen in 2017, by expanding its use in the private sector. One of the most promising areas of the space industry is the expansion of the use of satellite data such as remote sensing. It is expected that new space data utilization services will be established by integrating with recent innovative technologies such as AI (Artificial Intelligence), Big Data, and IoT (Internet of Things). The use of space technology will become more accessible in the future as rockets and satellites become smaller and less expensive.

The Ministry of Economy, Trade and Industry (METI) is studying the release of government satellite data to the public, making it free of charge, and improving the environment for satellite data use from the perspective of expanding the use of space data. In the future, access to satellite data will become easier, industries can be created with less investment, and new business applications will be promoted.

5.3.2 *Use of Small Satellites*

Major changes are taking place in the field of space development and utilization. One of these changes is the increase in space development by the private sector, including venture companies. Until now, space development was mainly R & D led by national organizations, but we have entered an era in which the private sector develops and operates satellites and rockets with private investment, and the government purchases their services.

In particular, the use of nanosatellites is progressing worldwide. Conventional earth observation satellites weigh less than 100 kg, while large satellites weigh several thousand kilograms. Compared to conventional large satellites, nanosatellites are said to cost about one-hundredth of the cost and take about one-half the time to develop. Therefore, not only developed countries but also developing countries in Asia can own satellites, and nongovernmental companies and universities can also develop satellites. Fukui Prefecture has been attracting attention for its development of a prefectural satellite that uses technology from companies in the prefecture. In the United States, more than 100 venture companies are developing and utilizing nanosatellites, and nanosatellite development ventures are attracting attention in Japan as well.

5.3.3 *Satellite Constellation*

Various services are being developed using satellite data observed from satellites. In particular, the frequency of ground-based observations from space is dramatically increasing, and satellite data is now being treated as part of big data. This dramatic increase in observation frequency is supported by satellite constellations. Satellite constellation is a method to realize a single function or service by linking multiple satellites. In conventional satellite remote sensing, a single large observation satellite is used to observe everything, so the observation cycle is once every few weeks, and when weather conditions are taken into account, the actual observation of the earth's surface may only be possible once every few months. With satellite constellations, the observation frequency can be dramatically increased by creating a cooperative observation system with multiple satellites.

The use of such new space data is due to technological innovations that have made small satellites more powerful and less expensive. In Japan, AXELSPACE Corporation is building a satellite constellation of 50 nanosatellites (GRUS satellites) in orbit to enable daily observations around the world. As a leading overseas example, Planet, Inc. of the United States has realized a satellite constellation of more than 130 small satellites (weighing approximately 5 kg), which are constantly taking images of the global land area. The ground resolution of the images taken is 3 m, and the company has started a service that provides data within 4–12 h after the images are taken (Fig. 5.6).

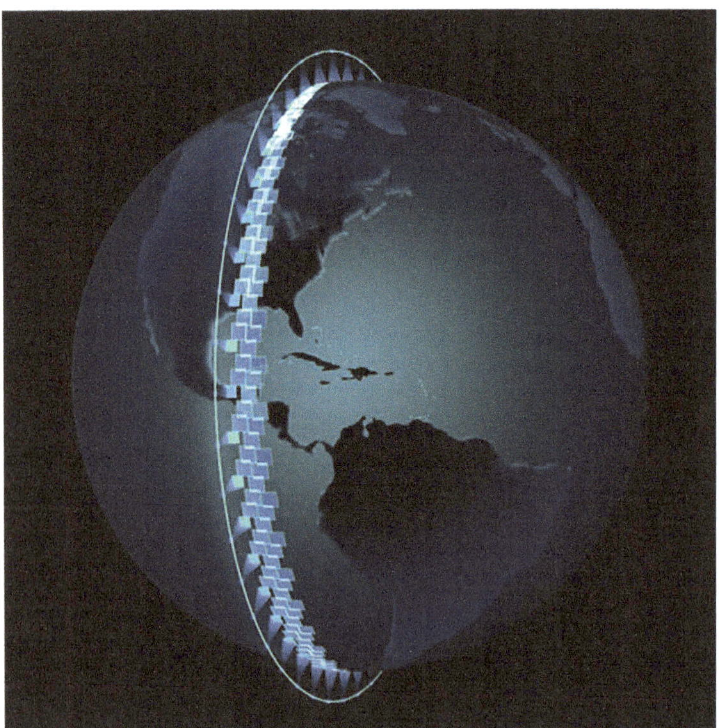

Fig. 5.6 High frequency observation by satellite constellation. (© AXELSPACE INC.)

5.4 Summary

In this chapter, the basic principles and applications of remote sensing using satellites are explained. It also explains the current status of space development and the future development of remote sensing technology.

Part III
Utilization of Geospatial Information in Daily Life

Chapter 6
Applications in Urban Facility and Land Management: From Urban Facility Management to Smart Cities

Yoshihide Sekimoto

Abstract There are various things that make up a city, including roads, underground structures, buildings, and land, and there are various ways to develop and manage the data, but we will deal with geospatial information comprehensively from the perspective of various urban plans, from routine to atypical, to get a bird's eye view of the city.

Keywords Urban facility information · Building information · Land use information · Urban planning · Compact city · Smart city

6.1 Urban Facility Management Starting from Underground Facilities

6.1.1 History Starting from Gas Explosion

When did geospatial information start to be used for the management of urban facilities? In the case of Japan, the Tenroku gas explosion in Osaka in 1970 may be one example. In the Tenroku gas explosion, a joint in an underground gas pipe near Tenjinbashi-suji 6-chome in Osaka came loose and city gas spewed out, igniting a huge explosion that killed 79 people and seriously injured 420 others. In the process, the system gradually evolved from an understanding of the current pipelines to the planned pipeline locations and priority sales locations with each household's potential customers in mind.

Today, water supply, sewerage, power and communication cables are similarly managed by their respective business entities using some kind of GIS and are also beginning to be managed on a common infrastructure (for example, the road

Y. Sekimoto (✉)
Center for Spatial Information Science, University of Tokyo, Tokyo, Japan
e-mail: sekimoto@iis.u-tokyo.ac.jp

Y. Sekimoto, Y. Kawahara (eds.), *Utilization of Geospatial Information in Daily Life*, New Frontiers in Regional Science: Asian Perspectives 65,
https://doi.org/10.1007/978-981-19-6905-8_6

management system by the Road Management Center) (http://www.roadic.or.jp/systemimage.html).

6.1.2 Road Management Information

Then, what about the management of the road facilities that stretch all over the country? Roughly speaking, there are about 8000 km of expressways, 20,000 km of national roads under the direct jurisdiction of the national government, 160,000 km of national and prefectural roads managed by prefectures, and 1 million km of municipal roads managed by municipalities. In addition to constructing new roads, the road managers at each site are required to record their work in the road ledger, which is required by Article 28 of the Road Law, for daily road management, such as filling potholes, collecting garbage, and removing snow. In addition, the Road Law Enforcement Regulations require the maintenance of detailed drawings called "road ledger charts." Despite the fact that it is essential for local governments to maintain this information in order to receive tax subsidies, there has not been much progress in digitization, and in the case of road ledgers, the rate was 50.1% as of 2012 (Ministry of Internal Affairs and Communications 2012).

On the other hand, from the usage side, car navigation systems were the first to be used when talking about roads, and after Honda released the world's first car navigation system, called the Electro gyrocator (Honda n.d.), in 1981, it became so popular that GIS was commonly referred to as car navigation. Since then, not only the private sector but also the public side has been developing Digital Road Map (DRM) in the 1990s. In the 1990s, the Digital Road Map (DRM) was developed, and VICS (Vehicle Information and Communication System) was used to distribute information on construction traffic regulations and traffic congestion, as well as to actively provide information on various road administrations, such as road traffic census surveys.

In addition, since the above data is rather for business use and is expensive, it was difficult to use it for information dissemination from the general public, so recently, the Open Street Map (OSM), which is developed and released free of charge, is increasingly used as a base (see Chap. 14 for details).

6.1.3 Data on Various Urban Facilities

So far, we have discussed underground utilities such as gas, electricity, and telecommunications, as well as roads, but what other data is available on urban facilities? The Ministry of Land, Infrastructure, Transport and Tourism's National Land Survey (http://nlftp.mlit.go.jp/ksj/), for example, provides information on public offices, schools, hospitals, post offices, and social welfare facilities in the category of public facilities, as well as the locations, types, and names of bus stops, railroads,

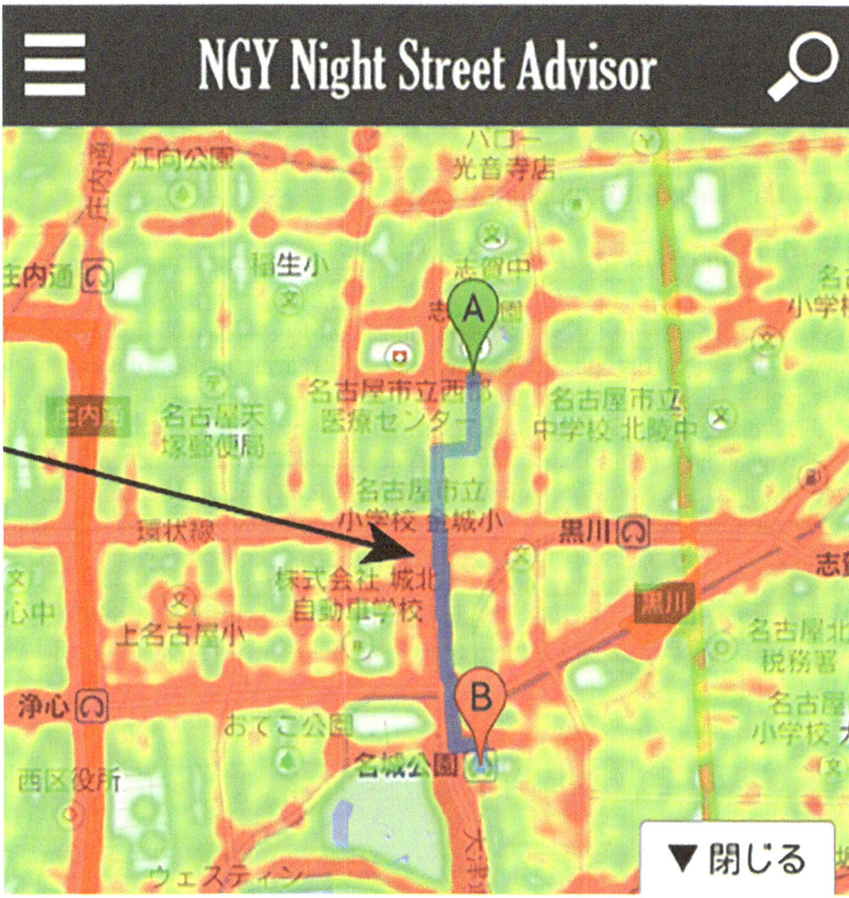

Fig. 6.1 Example of using lighting data. (Cited from Matsuda's document: http://i.csis.u-tokyo.ac.jp/event/20140623/index.files/140623_csisi_08_05.pdf)

airports, ports, and fishing ports in the category of transportation facilities. In the category of transportation facilities, the locations, types, and names of bus stops, railroads, airports, ports, and fishing ports across the country are maintained and disclosed, which is very valuable as a national database.

But what about more detailed data, such as the location of each lighting pole? In general, municipalities maintain lighting registers as part of their road facilities, but it seems that not many of them have the data down to latitude and longitude. However, with the recent spread of smartphones, it has become easier to create useful applications that can be easily understood directly by citizens, and one idea can encourage the creation of data. Figure 6.1 shows a smartphone application called Night Street Advisor, which was created by a student of Akashi National College of Technology. This application shows the brightness of the street for people walking at night, so that they can reach their destination through the brightest street possible. Naturally, this application requires data on each and every lighting pillar, but the city of Nagoya agreed with the purpose of the application and provided us with tens of thousands of lighting pillars to create this application.

6.2 To Grasp Buildings and Land

6.2.1 Understanding Property Taxes

In this section, we will focus on the buildings and land that make up a city. For municipalities that run cities, property tax is a large percentage of local taxes. Specifically, buildings and land are assessed (reassessed) once every 3 years based on the property assessment standards set by the Minister of Internal Affairs and Communications, and the taxable amount is determined. Under such circumstances, from the local government's point of view, it is important to keep track of the changes in the number of houses because in many cities, there are almost as many buildings as there are people. In many, but not all, cities, aerial photographs are used to record unassessed houses, unidentified houses, and houses whose area has changed (Fig. 6.2). Also, in 2007, Tondabayashi City discovered about 500 property tax leaks using a map information system based on aerial photographs, which

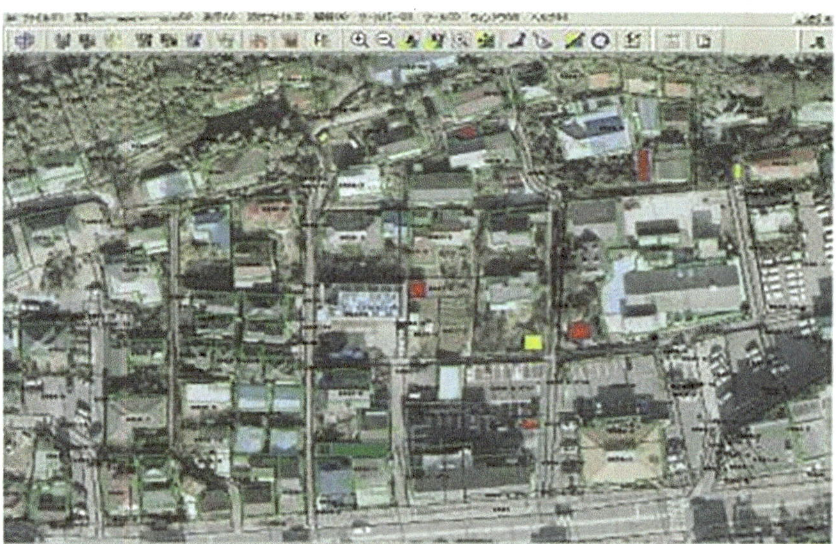

Fig. 6.2 Identification of houses subject to property taxation using aerial photographs. Red building means it is not evaluated and has no taxation. Yellow means it has taxation but is not identified. Green means it has gap between taxation area and actual. (Cited from http://www.recpas.or.jp *Not currently available)

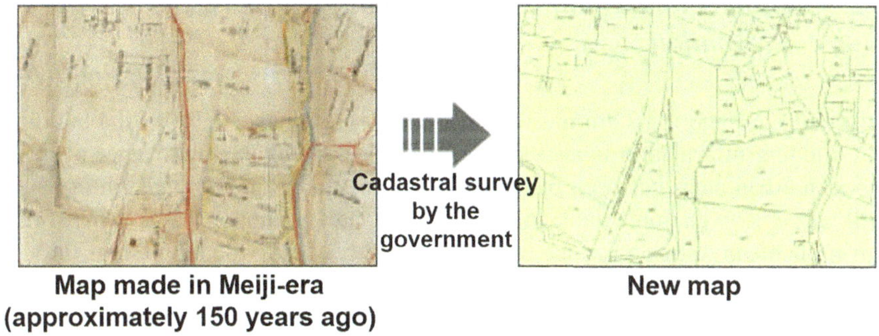

Fig. 6.3 A cadastral map updated from the official map by the cadastral survey. (Cited from Ministry of Land, Infrastructure, Transport and Tourism n.d.)

resulted in tens of millions of yen in additional taxation (http://www.gamenews.ne.jp/archives/2007/10/_500.html n.d.).

Specifically, the owner, lot number, and lot name of each parcel (one stroke) are surveyed, and the location and area of the boundary are surveyed, which is the so-called "family register" survey of land. About half of them are based on old maps (public maps) made at the time of land tax revision in the Meiji era (1868–1912), which are sometimes inaccurate. Therefore, a land registry survey is conducted, and the results are sent to the registry office to provide basic information for calculating property taxes (Fig. 6.3) (Ministry of Land, Infrastructure, Transport and Tourism n.d.).

6.2.2 *For Transparency and Revitalization of Real Estate Transactions*

While Sect. 6.2.1 was mainly about the shape of buildings and land, the actual market price at which they are bought and sold is also important for taxation purposes. In this case, information on when, where, and at what price the transaction took place is important. Currently, the Ministry of Land, Infrastructure, Transport and Tourism (MLIT) publishes information on real estate transaction prices together with maps in its Comprehensive Land Information System (http://www.land.mlit.go.jp/webland/). This system integrally publishes data on tens of thousands of national official land prices, standard land prices of prefectures, and questionnaires (about 2 million) based on the Real Estate Transaction Price Information System (Ministry of Land, Infrastructure, Transport and Tourism, Bureau of Land, Infrastructure, Transport and Tourism 2014).

6.2.3 Understanding the Age of Buildings for Disaster Prevention

Sections 6.2.1 and 6.2.2 are mostly related to the real estate value of the building, but building information is also key for disaster prevention. The danger of dense wooden urban areas has been mentioned for some time, but geospatial information will make it even more real. Although there are not many cases where local governments investigate the age, material, use, etc. of individual buildings through basic urban planning surveys, it is possible to estimate the risk (total destruction rate) for each area by combining earthquake simulations with some estimated values. Figure 6.4 shows the "Building Danger Map of Toyama City," which calculates the total destruction rate in the event of a large-scale earthquake based on the building conditions as of January 2010, the structure (wooden/non-wooden) and year of construction of buildings in each area, and the magnitude of shaking at each location.

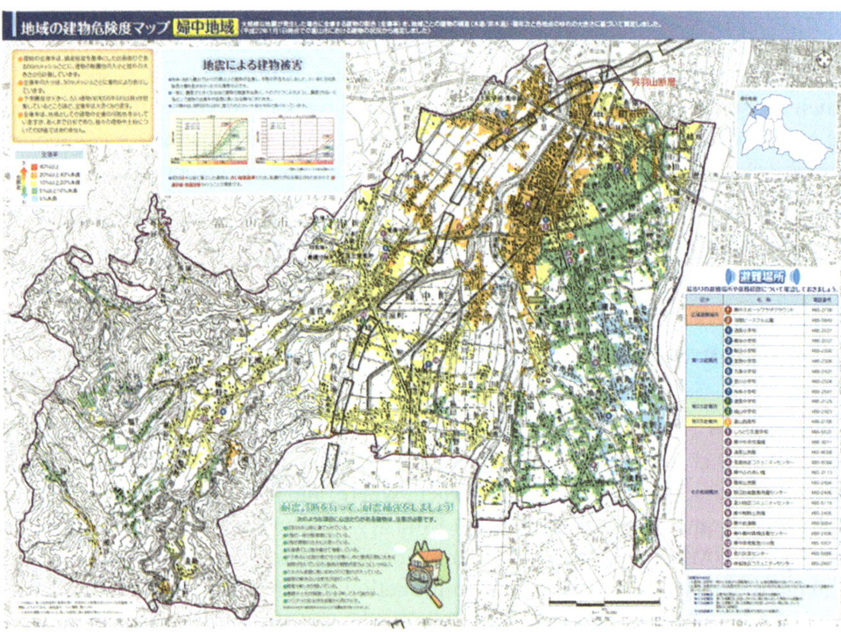

Fig. 6.4 Map of danger levels based on building age (Toyama City). Orange color means 20–40% of buildings will have full damage. Yellow means 10–20% and green means 5–10%. (Cited from https://www.city.toyama.toyama.jp/data/open/cnt/3/2674/1/bousaimappu_hutyuu2.pdf?20220329010319)

6.3 For Urban Planning

6.3.1 Basic Urban Planning Survey

So far, we have discussed individual urban facilities, buildings, land, and other major components, but how can geospatial information be used in the planning and evaluation of an entire city? For example, the Basic Survey on Urban Planning based on Article 6 of the City Planning Law is conducted approximately once every 5 years, and as shown in Table 6.1, the survey covers items related to population, industry, urban facilities, transportation, environment, disasters, and landscape. This information serves as the basis for various urban plans, and results in various regulations such as zoning, etc., which are reflected in city planning maps and other drawings (Fig. 6.5).

6.3.2 For Solving Various Urban Issues

In addition to the routine surveys and data development described in (1) above, recent years have seen the formulation of urban reconstruction plans in the wake of

Table 6.1 Items of urban planning basic survey (Ministry of Land, Infrastructure, Transport and Tourism, Urban Affairs Bureau 2013)

Type	Data item
Population	Population size, DID (densely inhabited district), future population, population change, commute migration, daytime population
Industry	Number of workers, number of establishments, number of employees, and sales value by industry and occupational category
Land use	Zoning status, land use status, national and public land status, residential land development status, agricultural land conversion status, forest land conversion status, new construction trends, ordinances and agreements, and application status of agriculture, forestry, and fishery-related policies
Building	Current status of building use, location of large retail stores, etc., and number of households by ownership relationship and standing of houses
Urban facility	Location and description of urban facilities, etc., and road conditions
Traffic	Cross-sectional traffic, congestion, and travel speeds on major trunk lines, automobile flow, railroads, streetcars, etc., and buses
Land price	Land price situation
Green environments	Topography, water system, geological conditions, weather conditions, green conditions, recreational facilities, flora and fauna survey
Disasters and pollutions	Occurrence of disasters, disaster prevention centers and evacuation sites, and pollution
Landscape and historical heritage	Status of tourism, landscape and historical resources, etc.

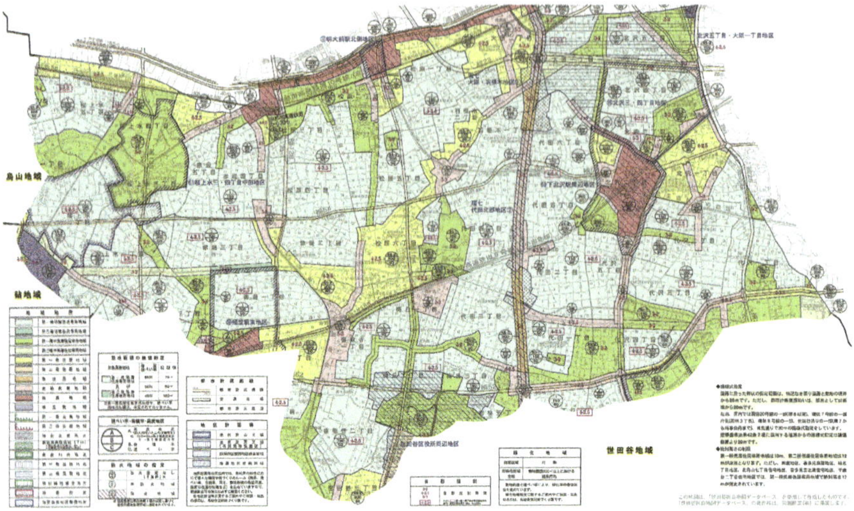

Fig. 6.5 Setagaya Ward City Planning Map (Kitazawa District): Red color means commercial zone. Yellow means category 1 residential district. Green means category 1 medium-to-high-rise exclusive residential district. Light blue means category 1 low-rise exclusive residential district. (Cited from https://www.city.setagaya.lg.jp/mokuji/sumai/001/001/d00004878_d/fil/1kitasawa.pdf)

the Great East Japan Earthquake, and responses to the problems of compact cities and empty houses in the face of declining birthrates and declining populations. It seems necessary that the use of geospatial information should also be diverse and flexible.

In the aftermath of the Great East Japan Earthquake, the Urban Planning Bureau of the Ministry of Land, Infrastructure, Transport and Tourism conducted the "Great East Japan Earthquake Tsunami Affected Urban Area Reconstruction Support Survey" on behalf of all the affected municipalities in the Tohoku region, and formulated reconstruction plans for each municipality. These surveys are archived in collaboration with the University of Tokyo's Research Center for Spatial Information Science (http://fukkou.csis.u-tokyo.ac.jp) and are being made available to the public (see Sekimoto et al. 2013). Details are described in Chap. 10.

From the aspect of compact cities, it has also become necessary to consider streamlining and networking of existing public facilities and residential areas. For example, in Fig. 6.6, in 2009, Niigata City prioritized future facility development based on population density and the layout of existing community facilities, dividing them into districts that should be consolidated or eliminated, districts that should be maintained, and districts that should be newly constructed, and displaying them on a map. Such efforts have also begun in Japan, where the Law on Special Measures for Urban Renewal was partially revised in August 2014, and the *Handbook on the Evaluation of Urban Structures* was published. In this handbook, methods for evaluating accessibility to various facility layouts are described.

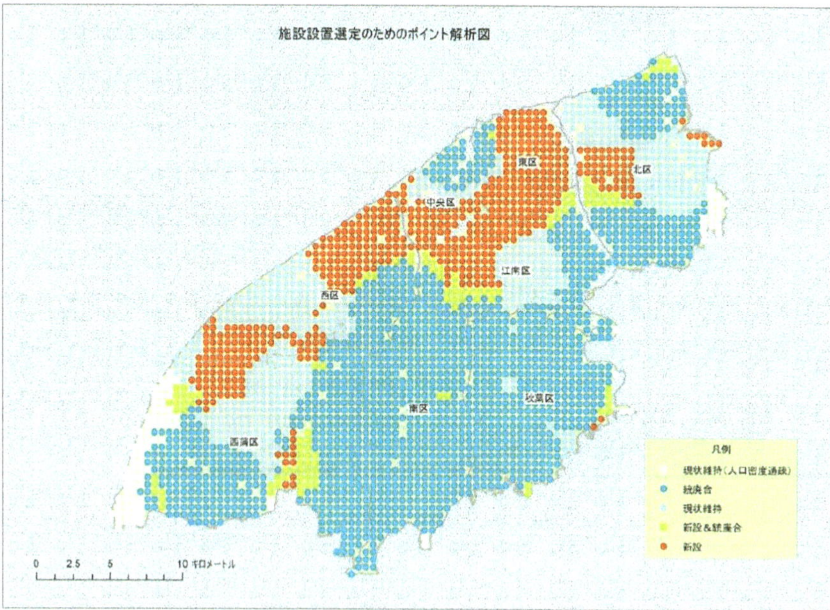

Fig. 6.6 Prioritization of future facility development based on population density and existing community facility layout in Niigata City. Red color means new development, and light blue means consolidation and abolition. Gray means maintenance of the current status. (Cited from http://www.city.niigata.lg.jp/shisei/soshiki/soshikiinfo/toshiseisaku/gis.files/v3_2_110521.pdf)

6.3.3 *Sharing City Data Between Citizens and Officials*

If we think about it a little more closely, is city planning something that the people of the local government are fully responsible for? In recent years, there has been an increase in the number of projects that are not left up to the government, but are considered by the citizens themselves. For example, the CityDashboard in London, which was launched by UCL (University College London) in 2012, is based on data that is already available to the public and also includes links to image data from cameras on the streets provided by the government, allowing users to get a literal overview of the city. In this way, it is possible to get a literal overview of the city. This kind of citizen participation using open data, which will be discussed in detail in Chap. 14, has the potential to increase and become mainstream as a new approach that has the freedom to change the way cities are organized.

In Japan, for example, the My City Forecast (https://mycityforecast.net) shows the long-term future of a region, which is difficult for local authorities to present, on a detailed and intuitive scale so that citizens can easily have a sense of ownership (Fig. 6.7) (see Hasegawa et al. 2019). It compares how much the accessibility to various facilities and the cost burden for maintaining facilities differ depending on whether or not the area represented by a 500 m mesh has a location optimization plan

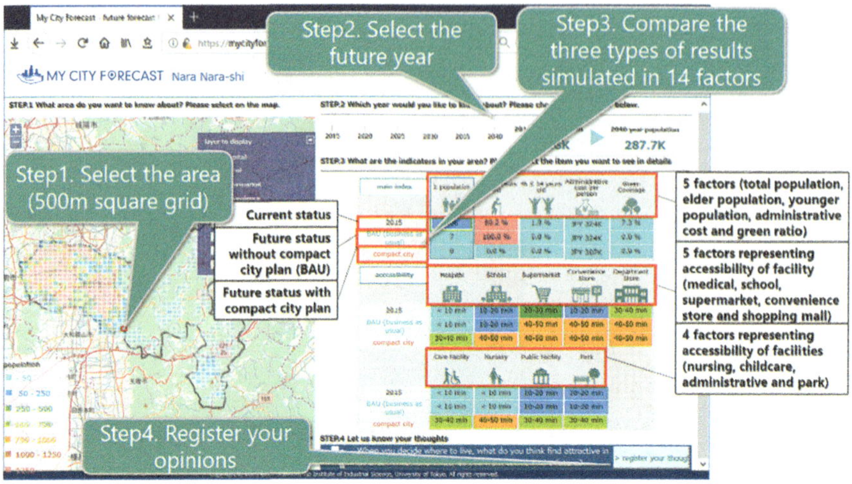

Fig. 6.7 My City Forecast, a detailed scale representation of the future vision of the region. (The site is an example from Nara City. https://mycityforecast.net/)

defined in the compact city plan at present (2015 in the screen) and in the future (2040 in this case). Specifically, the future population distribution is estimated using the national census, future population projection data, and building data that are open to each municipality. Finally, administrative costs and accessibility to facilities are calculated using the population distribution and location of facilities, as well as municipal financial information and public transportation information. This makes citizens more aware of their role in the process.

In addition, in recent years, there has been an increasing number of efforts worldwide to represent cities more realistically in three dimensions. In Japan, PLATEAU (https://www.mlit.go.jp/plateau/), led by the Urban Affairs Bureau of the Ministry of Land, Infrastructure, Transport and Tourism, was released in December 2020. By effectively utilizing the two-dimensional polygons, height information, and other attribute information of building information described so far in this chapter, it is now possible to have a bird's eye view of a three-dimensional city model on a browser (Fig. 6.8). However, as more accurate data is required, maintaining and updating the data in a sustainable manner will be an important issue in the future.

6.4 Conclusion

In this chapter, we have taken a bird's eye view of the information that deals with the various components that surround a city and have discussed how geospatial information has been handled in this field so far and how it will be developed in the future.

Fig. 6.8 Three-dimensional digital twin PLATEAU released by the Ministry of Land, Infrastructure, Transport and Tourism. Chiyoda, Chuo, Minato, Shinjuku, and Bunkyo wards are color-coded by building height. (Cited from https://www.mlit.go.jp/toshi/city_plan/content/001388017.pdf)

References

Hasegawa Y, Sekimoto Y, Seto T, Fukushima Y, Maeda M (2019) My city forecast: urban planning communication tool for citizen with national open data, computers, environment and urban systems, vol 77. Elsevier, Amsterdam, p 101255

Honda (n.d.) History of car navigation. http://world.honda.com/history/challenge/1981navigationsystem/index.html

http://www.gamenews.ne.jp/archives/2007/10/_500.html (n.d.)

Ministry of Internal Affairs and Communications (2012) Local government information management overview

Ministry of Land, Infrastructure, Transport and Tourism (n.d.). Land registry survey. http://www.chiseki.go.jp

Ministry of Land, Infrastructure, Transport and Tourism, Bureau of Land, Infrastructure, Transport and Tourism (2014) Basic concept of information stock system related to real estate

Ministry of Land, Infrastructure, Transport and Tourism, Urban Affairs Bureau (2013) Urban planning basic survey implementation guidelines

Sekimoto Y, Nishizawa A, Yamada H, Shibasaki R, Kumagai J, Kashiyama T, Sagara T, Kayama Y, Otomo S (2013) Promotion of data distribution by building a survey archive for the Great East Japan Earthquake reconstruction support. GIS Theory Appl 21(2):1–9

Chapter 7
Applications in Transportation Systems and Mobile Devices: Capturing the Movement of People

Yoshihide Sekimoto

Abstract This chapter provides an overview of the technologies and their evolution in capturing people's movements by using data from transportation systems and mobile terminals. It also provides an overview of the handling of personal information surrounding location information and the future of such information.

Keywords Transportation · Mobile devices · GPS · Mobile terminals · Smartphones · Personal information · Behavioral change

7.1 Use in Transportation Systems

7.1.1 Use in Road Transportation Systems

In this section, we will first touch on the road traffic system. In Japan, the ETC (Electric Toll Collection) system for automatic toll collection was researched and developed mainly by the government in the 1990s and was put into general use on expressways nationwide in November 2001. The system uses the Dedicated Short-Range Communication (DSRC) method so that the ETC on-board unit in the vehicle can communicate with the antenna on the roadside in both directions wirelessly in a short time of about 1–2 s (Fig. 7.1). This system has drastically changed the way traffic congestion at toll gates is greatly reduced and expressway tolls can be flexibly changed according to date, time, and vehicle type. The Ministry of Land, Infrastructure, Transport and Tourism (MLIT) has reported that as of October 2020, there are 138 ETC-only interchanges nationwide.

In addition, at the same time, Japan also developed the Vehicle Information and Communication System (VICS), a system that measures passing vehicles based on beacon receivers installed in the car and displays traffic congestion on the car

Y. Sekimoto (✉)
Center for Spatial Information Science, University of Tokyo, Tokyo, Japan
e-mail: sekimoto@iis.u-tokyo.ac.jp

Y. Sekimoto, Y. Kawahara (eds.), *Utilization of Geospatial Information in Daily Life*, New Frontiers in Regional Science: Asian Perspectives 65,
https://doi.org/10.1007/978-981-19-6905-8_7

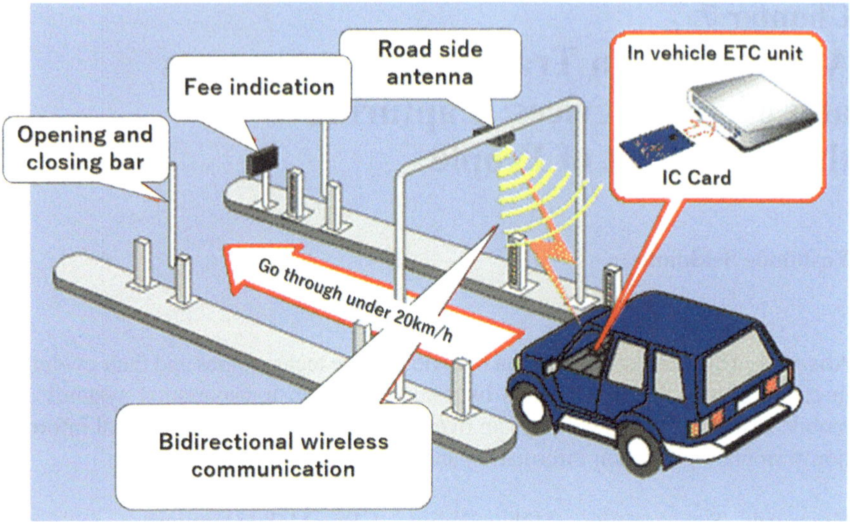

Fig. 7.1 Measurement of road traffic volume by ETC. (Created based on https://www.its-tea.or.jp/its_etc/service_etc.html)

navigation system. The ETC2.0 system has the function of a probe car and can be installed at 1700 ITS spots nationwide on the roadside of expressways. ETC2.0 has the function of a probe car and is uploaded from 1700 ITS spots, which are installed on the roadsides of expressways and other locations throughout Japan. Many countries around the world have introduced DSRC-based systems for automatic toll collection on expressways.

7.1.2 Use in Railroad Systems

Next, I would like to discuss railroad systems. It is noteworthy that JR East developed the Suica automatic ticket gate system in the 1990s, at about the same time as the aforementioned ETC system, and launched the service in November 2001. In the initial testing stage, Suica used a quasi-microwave communication method that is strong in the height direction, assuming that the user would hold it over the automatic ticket gate. It is interesting to note that the touch-and-go system was established by adopting shortwave technology, which is strong in the horizontal direction, and changing to the "touch" method.

In addition, each automatic ticket gate is managed by a management server in each station, and the data at each station is managed autonomously and decentralized, and then aggregated by the national server. This system is designed to manage risk so that even if there is a problem at one of the automatic ticket gates, it will not immediately affect the entire country. In fact, in 2006–2007, some manufacturers' automatic ticket gates had trouble due to bugs, but train operations were not affected.

It is also interesting to note that in order to introduce a large-scale system like Suica at that time, the company explained that the initial investment would pay for itself in 10 years through the reduction of paper and ink due to the reduction of paper tickets and the reduction of staff response time due to ticket gate problems. Although the company has since made great strides in electronic payment with Suica and the station shop business, the fact that it has not used these factors as a persuasive factor is significant, and may be a device for new investment within the organization (Shiihashi n.d.; Urihara n.d.; East Japan Railway Company HP n.d.).

Overseas, IC card-based transportation systems have been widely adopted, but at the same time, the systems themselves can become complicated, so some countries have designed their systems to avoid the introduction of automatic ticket gates by strictly enforcing fines.

7.1.3 Use in Bus Systems

Compared to roads and railroads, buses are operated by individual bus companies or local governments on a smaller scale, and therefore tend to be slow to adopt IT. Although there are more than 2000 bus companies in Japan that provide integrated transfer information services for rail and bus based on private-sector timetables, the problem has been that the bus companies themselves do not necessarily have easy-to-use data. In recent years, however, efforts have been made to standardize bus information across the country using the General Transit Feed Specification (GTFS), which is being developed and opened up at the grassroots level.

In Japan, the Ministry of Land, Infrastructure, Transport and Tourism (MLIT) established the Japanese version of GTFS, GTFS-JP, in March 2017. In Japan, MLIT established GTFS-JP as the Japanese version of GTFS in March 2017, and as of December 2020, 289 companies have developed and released data in accordance with this standardization (Fig. 7.2). However, this is not the only advantage. Standardization also reduces the cost of building various systems such as timetabling tools and digital signage.

As mentioned above, bus operations are often more financially difficult in regional cities, so it is hoped that such standardization will allow for the common use of tools, low-cost IT, and efficient operations in the future.

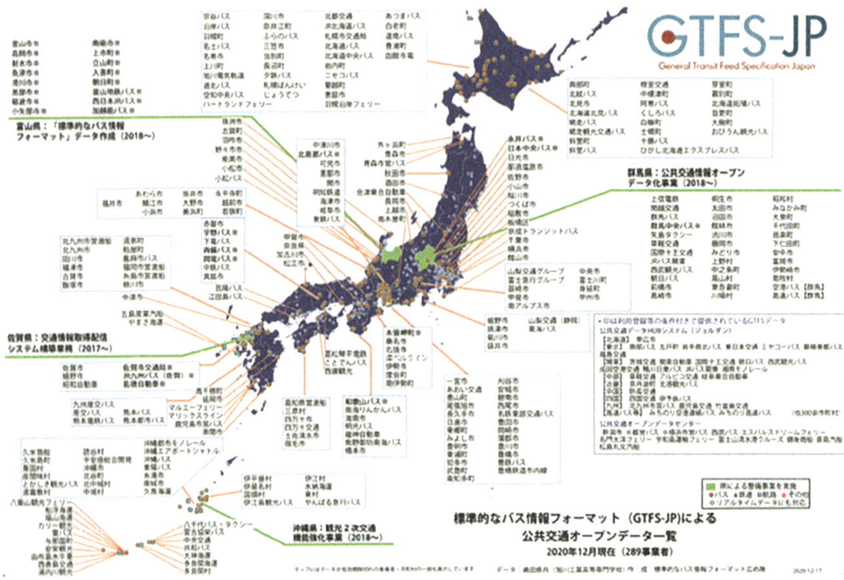

Fig. 7.2 Spread of bus open data by GTFS. (Cited from https://gtfs.jp/blog/)

7.2 Use in the Flow of People

7.2.1 Reproduction of People Flow in Urban Areas Using Person Trip Survey Data

In this section, I would like to discuss the technology for comprehensively grasping the movement of people, while the use in the transportation system in Sect. 7.1 was mainly about how to count the number of people for each transportation method. In particular, the flow of people is expressed as a moving object that contains time information, which is the spatio-temporal data described in Chap. 3. How to acquire the data is the key, but first, I would like to describe a questionnaire survey called a person trip survey. As shown in Fig. 7.3, it is a paper-based survey that describes the daily activities of a typical weekday or holiday during the survey period, mainly several destinations and the movements between them.

Unlike GPS data, which will be discussed later, this data only provides the approximate name, time, and means of transportation for each destination, but it is very valuable as an unbiased public survey because the sample rate is only a few percent of the total population of the metropolitan area. Sekimoto et al. converted these data into latitude and longitude, interpolated the shortest route between the start and end points of each trip (see (Sekimoto et al. 2011a) for details), and provided the data for public use in the "People Flow Project" (https://pflow.csis.u-tokyo.ac.jp/) in the Center for Spatial Information Science at the University of

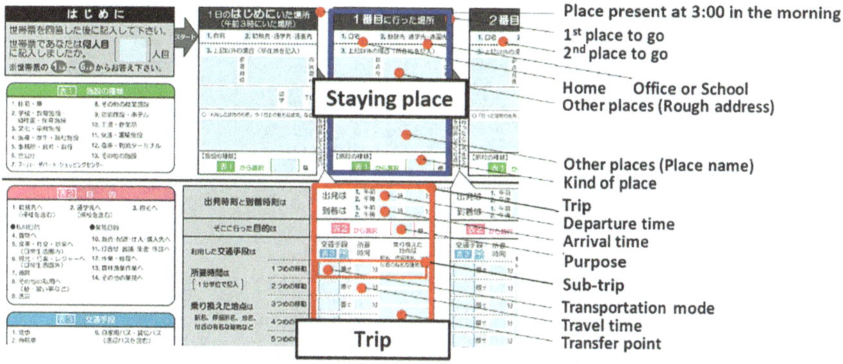

Fig. 7.3 Questionnaire for the person trip survey. (Created based on Sekimoto et al. 2011a)

Tokyo. As of 2020, it has provided data for a total of seven million people in 36 metropolitan areas in Japan and overseas.

7.2.2 *Utilization of GPS and Base Station Usage Data of Mobile Terminals*

With the advent of the smartphone era in the late 2000s, GPS chips have become a standard feature in mobile terminals, making it possible for each device to acquire its own location information. For example, at the time of the Great East Japan Earthquake in 2011, through joint research with cell phone companies, we can see how people's behavior changed rapidly before and after the earthquake (Fig. 7.4).

However, the collection of location information obtained from individual mobile terminals itself, as will be discussed in more detail next, is being handled with caution due to the growing awareness of personal information protection in recent years. In this sense, the Call Detail Record (CDR) differs from GPS in that it is based

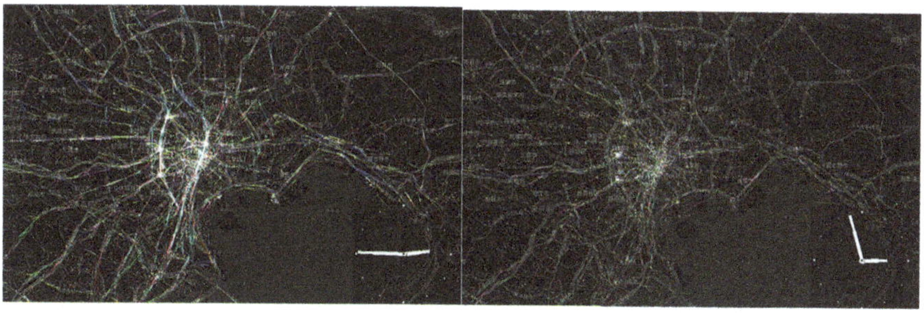

Fig. 7.4 Flow of people in the Tokyo metropolitan area before and after the Great East Japan Earthquake. (Cited from Sekimoto 2013)

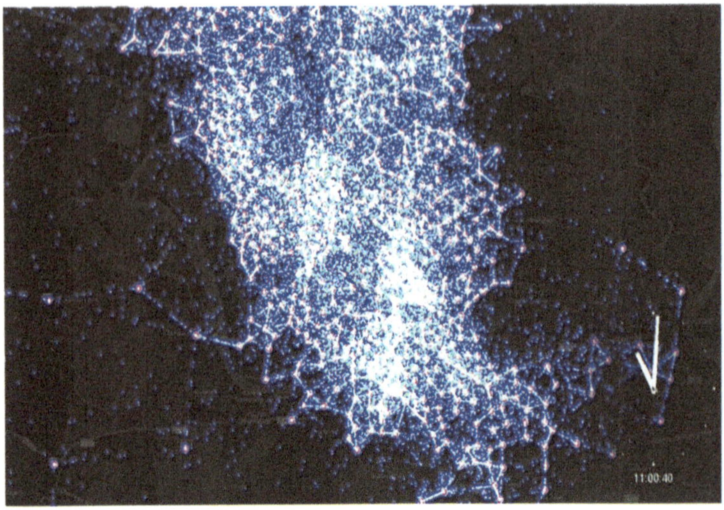

Fig. 7.5 People flow in Yangon using CDR

on the location information of the base station accessed during communication and the start/end time data of the communication. Since base stations of mobile terminals exist at intervals of several hundred meters to several kilometers, depending on the population density, the location of each individual is not accurately represented in this sense, but the number of people can be determined to some extent by aggregating the data using mesh, etc., and this data is being commercialized as private data such as mobile spatial statistics. In developing countries, the use of CDR is also effective because it eliminates the need to install new sensors.

In particular, Fig. 7.5 shows the flow of people in Yangon, Myanmar, using CDRs (see Sekimoto et al. 2015 for detail method). However, CDRs have the advantage of having a higher sample rate than GPS because they serve as a digital footprint and cover basically all terminals in the vicinity of the base station at the time recorded by the CDR.

7.2.3 The Suica Incident and the Concept of Personal Information

However, there were not all bright sides. From the perspective of a citizen, some of the information falls under the category of personal information, and there may be some concerns about privacy. In recent years, big data has been gaining momentum, and in May 2013, Prime Minister Abe's New Growth Strategy speech included the phrase "GPS is a treasure trove." The current Personal Information Protection Law does not allow businesses that have obtained permission from individuals to use their personal information to provide it to third parties, so there is a gap between the desire for smooth use from a business perspective and the lack of legislation to bridge the gap and promote safe data use.

To compensate for this, the "Study Group on Personal Data" (chaired by Masao Horibe, Professor Emeritus of Hitotsubashi University) of the government's IT Strategic Headquarters (headed by Prime Minister Abe) started discussions in September 2013, and as a review in line with the actual situation of personal data utilization in the era of Big Data, the group decided to "reduce the possibility of personal identification to a certain level" and to "establish a system to ensure that personal data can be used safely". While allowing flexible handling of personal data, such as providing it to a third party without the consent of the individual, it also "establishes statutory obligations to be assumed by business operators". The bill also clarifies the requirements for exceptions to the provision of personal data to third parties, such as joint use and opt-out, and establishes procedures to enable the ex-post expansion of the purpose of use.

7.2.4 *Positioning of Location Information*

For example, the "Commentary on the Guidelines for Personal Information Protection in the Telecommunications Business (Final Amendment: Notification No. 276 of the Ministry of Internal Affairs and Communications in 2010)" states that location information of cell phones is considered to be a matter that should be protected as privacy. It is considered to be a matter that should be protected as privacy. (Details are also summarized in Sekimoto et al. 2011b.)

On the other hand, what does it mean that an individual can be identified from location information? Technically speaking, it means whether or not a person can be identified as a specific person using some fragmentary location information including time as a clue. The paper "Unique in the crowd" by Y. A. de Montjoye et al. published in Nature in 2013 shows how the spatial and temporal resolution of the location information can be determined using base station-based mobile usage data (de Montjoye et al. 2013). In their paper, "Unique in the crowd", A. de Montjoye et al. use base station-based mobile phone usage data to calculate the probability that the trajectory reproduced from the data is unique, depending on the spatial and temporal resolution of the location information (de Montjoye et al. 2013). Specifically, Fig. 7.6 shows the probability of identification according to the time resolution on the horizontal axis and the spatial resolution (the degree of aggregation of base stations) on the vertical axis in the case where the number of points of cell phone usage data acquired by an individual is 4 on the left and 10 on the right. Of course, the higher the resolution of both time and space, the higher the probability of being identified, but it also shows that the probability is higher when the resolution of either time or space is quite high, rather than when the resolution of both is only moderate. As can be seen from the left side of Fig. 7.6, even if there are only four points, for example, a base station antenna with a temporal resolution of 3 h and a spatial resolution of three antennas has a probability of being identified of 0.7.

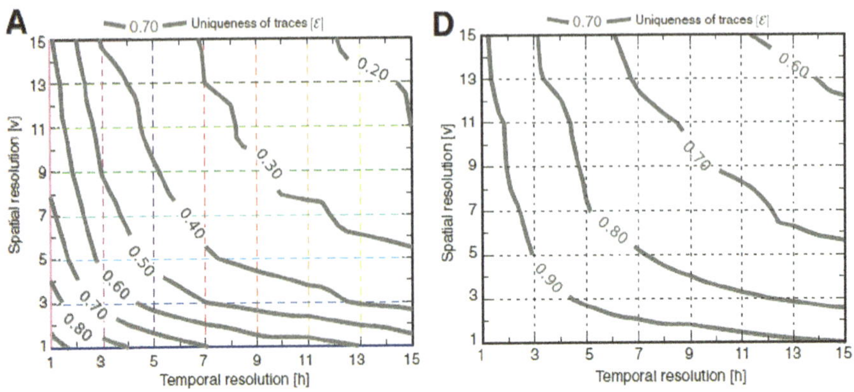

Fig. 7.6 Uniqueness of the trajectory calculated according to the number of location data that make up the trajectory. (Cited from de Montjoye et al. 2013.) The horizontal axis is the time resolution, which is approximately the time interval of acquisition. The vertical axis is the spatial resolution, which indicates the degree of aggregation of the antennas of the base station. (On the left is the case with 4 points of location data, and on the right is the case with 10 points)

7.2.5 For Proactive Use

While the previous discussion is essential for the safe use of personal information, isn't it also very time-consuming for you to actively use your own information? For example, if you enter your ID in some service, you can see your usage status, but there are not many such services, and it is not possible to see more detailed information such as today's physical condition and health status, or cell phone usage status including location information. It is also a hassle to have to enter your own information for each service. In response to this situation, an initiative called the "information bank" has been launched to allow people to actively use their own information, manage their own personal information as their own responsibility, and connect it to various services.

7.3 Future People Mobility and Transportation Systems

7.3.1 MaaS System Using Smart Phones

In this section, we will look at the use of geospatial information in the future of people's mobility and transportation systems. In recent years, with the simultaneous spread of motorization and smartphones in the Asian region, applications for easily catching a cab have emerged. Here, I would like to introduce GrabTaxi Holdings, which was founded in 2012 and operates in 17 cities in six countries: Malaysia, the Philippines, Thailand, Singapore, Vietnam, and Indonesia.

The process is not that difficult. When you launch the application on your phone and enter your destination, it will display all the cabs in the vicinity along with your current location and give you an approximate fare. You can then click on the button to make a reservation, and a nearby driver will be selected to come to you. It is also possible to cancel the reservation or call the driver directly. The driver, on the other hand, can easily respond to the notification by holding up his phone in the car. In 2014, the Softbank Group invested 30 billion yen in the service, and it has become a hot topic. In 2014, SoftBank Group invested 30 billion yen in the company, and it is often seen as a way to get a safe cab ride in Asia.

In Europe and the U.S., Uber, a company that started in San Francisco in 2009, is well known, and in Japan, some services have started in urban centers, but they are still in a state of limbo, often struggling with consistency with taxi-related laws in other countries.

MaaS (Mobility as a Service) is an extension of car-dispatch apps such as Uber and Grab, and is a system that allows payment for public transportation using a smartphone app, and is expected to increase gradually.

7.3.2 *Automated Driving and High-Precision Maps*

On the other hand, when we look at automobiles themselves, the development of technology for automatic driving has advanced considerably. While much of the work is done by sensors such as radar, images, and GPS on the vehicle side, the extent to which map information is used to support this is a competitive area for each company.

In terms of road maps themselves, as described in Chap. 6, Japan led the way in the 1980s and 1990s, mainly in digital road map technology for car navigation systems, but since the 2000s, with the emergence of Google Maps and global navigation system manufacturers, Japan has been somewhat overwhelmed.

In this context, it is also important to note that in line with automated driving, what used to be called a map is now conceptualized as a dynamic map, which consists of various layers of data, such as static, semi-static, semi-dynamic, and dynamic information, and data development and provision according to needs are now considered (Fig. 7.7).

However, static and quasi-static information on roads themselves will continue to be universally important for road administrators to maintain roads that are safe for vehicles to drive on before automated driving.

For example, through the activity called My City Report for Road Managers, we are using deep learning technology to detect potholes and cracks on roads in real time from images obtained by road managers driving with their smartphones on the dashboard (see Maeda et al. 2018 for details), and the accumulation of such data is also important.

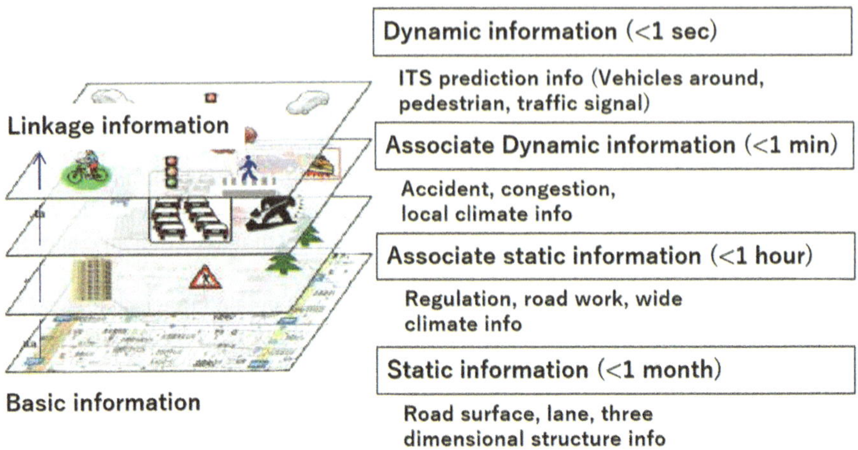

Fig. 7.7 Concept of dynamic map. (Created based on https://www8.cao.go.jp/cstp/gaiyo/sip/iinkai/jidousoukou.html)

7.3.3 People Mobility and Behavior Change

On the other hand, how will people's mobility change in the future? Until now, there has been a trend in Japan, where the population is expected to decline in the future, that people should live compactly, concentrating as much as possible in the central area of each region. However, in the wake of the corona disaster that changed the world drastically in 2020, the declaration of a state of emergency and the repeated waves of infection have created a trend toward dispersed and remote living and minimal encounters, while somehow maintaining economic activities.

In this sense, it will be important to continue to share information on the current situation and how to deal with it while using technology effectively in the rapidly changing environment. Figure 7.8 shows the relationship between the contact status of people in Tokyo in the months before and after the declaration of a state of emergency in the first half of 2020 based on GPS data from mobile terminals and the

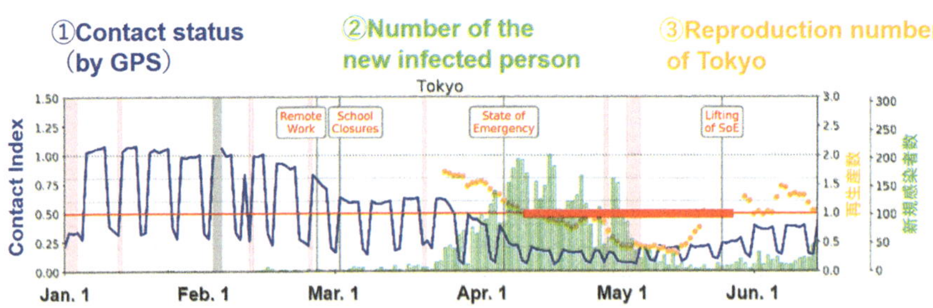

Fig. 7.8 Relationship between people's contact status and the number of reproductions before and after the declaration of a state of emergency. (Cited from Yabe et al. 2020)

status of infection through reproduction numbers. Specifically, the contact index is the ratio of the number of people who are in contact with each other when the contact status on weekdays during normal times before the corona disaster is set to 1, and the relationship between the contact index and the reproduction number is examined. The definition of contact itself is that two users outside their home are considered to be in contact if they stay within 100 m of each other for more than 30 min. When the number of reproduction exceeds 1, the number of infected people goes in the direction of increasing (red bold line area), and when we look at the relationship between that and the contact index, the boundary value is generally around 0.25–0.3. In other words, it is a good rule of thumb to keep contact levels to 1/3 to 1/4 of normal levels.

7.4 Conclusion

In this chapter, we started with the use of geospatial information in transportation systems and discussed the measurement of people's movement in general, its representation, and future prospects.

References

de Montjoye YA, Hidalgo CA, Verleysen M (2013) Unique in the crowd: the privacy bounds of human mobility. Sci Rep 3:1376

East Japan Railway Company HP (n.d.) The birth of Suica. http://www.jreast.co.jp/development/story/

Maeda H, Sekimoto Y, Seto T, Kashiyama T, Omata H (2018) Road damage detection and classification using deep neural networks with smartphone images. Comput Aided Civ Inf Eng 33(12):1127–1141

Sekimoto Y (2013) People flow and spatio-temporal dataset frontline. J Operat Res 58(1):24–29

Sekimoto Y, Shibasaki R, Kanasugi H, Usui T (2011a) Yasunobu Shimazaki PFLOW: reconstruction of people flow recycling large-scale social survey data. IEEE Pervasive Computing 10(4): 27–35

Sekimoto Y, Horanont T, Shibasaki R (2011b) Commentary: Current trends in people flow analysis technology using mobile phones. Inform Proc 52(12):1522–1530

Sekimoto Y, Kashiyama T, Hasegawa Y, Kanasugi Y (2015) Estimation of link traffic volume using sparse mobile phone call history—a case study in Dhaka. J Traf Eng 1(4):A_1–A_8

Shiihashi A (n.d.) The day penguins flew in the sky. Kotsu Shinbunsha Shinsho

Urihara S (n.d.) Outline and data utilization of suica. CSIS-i Open Symposium, University of Tokyo. http://i.csis.u-tokyo.ac.jp/event/20130129/index.files/130129_csisi_04_2-2.pdf

Yabe T, Tsubouchi K, Fujiwara N, Wada T, Sekimoto Y, Ukkusuri SV (2020) Non-compulsory measures sufficiently reduced human mobility in Tokyo during the COVID-19 epidemic. Sci Rep 10(1):18053

Chapter 8
Applications in Crime Prevention

Ikuho Yamada

Abstract Crime occurrence is not random, either spatially or temporally. It concentrates on a small number of specific locations or time periods; therefore, geospatial information is essential for crime research and prevention efforts. This chapter briefly introduces the history of the use of maps for crime analysis and then discusses the application of geospatial information for crime prevention and safety, beginning from the visualization of crime occurrence through to its prediction.

Keywords Crime mapping · Crime hot spots · Crime prevention efforts in local communities · Environmental factors related to crime occurrence · Crime causation theory and crime opportunity theory · Crime prevention systems

8.1 A Brief History of the Use of Maps in Crime Research

8.1.1 Crime Mapping

Crime does not occur randomly in space or time. Therefore, consideration of the spatial or temporal patterns of crime is necessary to effectively counter the risk and protect neighborhoods. Maps have been traditionally used to capture the spatial patterns of crime occurrence, and this visualization of crime occurrence in maps is known as crime mapping.

In the early nineteenth century, A.-M. Guerry, a French statistician, and L.A.J. Quetelet, a Belgian mathematician and statistician, created the earliest examples of crime maps, which formed the cartographic school of criminology. Those researchers both mapped crime occurrence rate by department in France and investigated its potential relationships with the socioeconomic characteristics of the regions. Figure 8.1 presents Guerry's maps showing the spatial distribution of

I. Yamada (✉)
Center for Spatial Information Science, The University of Tokyo, Chiba, Japan
e-mail: iku.yamada@csis.u-tokyo.ac.jp

Y. Sekimoto, Y. Kawahara (eds.), *Utilization of Geospatial Information in Daily Life*, New Frontiers in Regional Science: Asian Perspectives 65,
https://doi.org/10.1007/978-981-19-6905-8_8

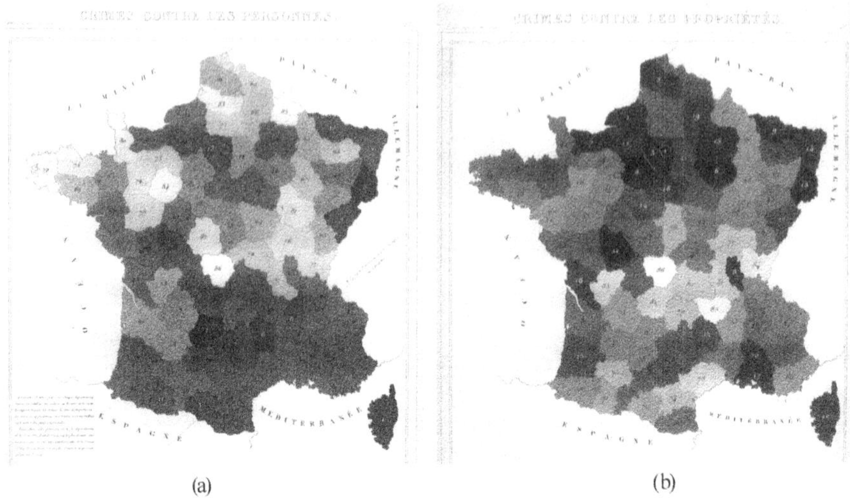

Fig. 8.1 Spatial distributions of crimes by department in France. (a) Personal crimes, (b) Property crimes. (From Guerry, A.-M. (1833). Essai sur la Statistique Morale de la France. Paris: Crochard)

personal and property crimes. Personal crime is perpetrated against a person, such as murder, assault, and robbery, and property crime is that against private properties, including such infractions as burglary, theft, and arson. The shading of each department in the figure indicates the population per crime incident. Darker shading corresponds to higher index values, indicating lower crime rates relative to the population. These maps suggest that crimes were unevenly distributed over space, and the spatial patterns of their distributions differed by crime category.

Crime research utilizing maps did not prosper during that period because of the lack of a theoretical basis to relate crime occurrence to regional characteristics and the limitations of analysis using paper maps. However, it began attracting research interest again in the early twentieth century through studies on regional risk factors of crime conducted by Chicago school researchers. Section 8.2.2 will discuss their work in detail.

8.1.2 Introduction of Computers into Crime Analysis

Crime analysis from a geographic perspective was aided by the use of computers beginning in the mid-1960s, although computers at that time were limited in capability for graphic presentation (Harada 2009). The computer maps created by the St. Louis Metropolitan Police Department in the U.S. for efficient patrolling are considered the first examples of such applications. Subsequent development of computers with graphic functionalities advanced crime mapping and computer-based crime analysis. For example, the Illinois Criminal Justice Information

Authority developed a computer program called the Spatio-Temporal Analysis of Crime (STAC) for clearer representation and more efficient analysis of crime occurrences (Bates 1987). It provided tools to analyze spatial and temporal patterns of crime occurrence, such as for determining crime hot spots that observed substantial concentrations of incidents.

GIS software for personal computers was first used in the 1990s, and it has helped researchers and practitioners in police departments use geospatial information to analyze and control crimes. In the U.S., the National Institute of Justice supervised demonstration experiments by police departments in five cities, such as Jersey City, where GIS was utilized to analyze the geographic distribution of drug-related crimes and develop controlling strategies and measures. Similarly, the New York City Police Department adopted a computer system called CompStat to analyze and map crime data to systematically manage police activity based on that data. The introduction of geographical analysis using GIS to the management of police activity is considered to have successfully reduced incidents at that time, leading to future developments of data-driven policing and geographic crime prediction.

8.1.3 Applications for Safety and Security of Neighborhood Communities

GIS and geospatial information are now widely used, and efforts have been made to expand their use to protect the safety and security of neighborhood communities. In Japan, researchers tend to take the initiative in such projects. This section introduces two examples related to the safety of children engaging in outdoor activities after school.

In the central part of Tokyo, members of a Parent-Teacher Association (PTA) and GIS researchers set up a project to assess and improve traffic safety of the routes to and from school (Imai 2015). They collected information on the locations where students and PTA members felt unsafe and summarized it using GIS maps. They also visited locations that many people considered unsafe to investigate the reasons for those opinions. Workshops were then held to discuss safety issues on the routes and possible solutions based on the GIS maps and results of the field survey. GIS is an effective tool for synthesizing and visualizing ambiguous information, such as the public conception of a location, to facilitate discussion among the people concerned.

Amemiya et al. (2009) used small GPS loggers to investigate after-school activities and safety of children. Those authors asked elementary school children from a particular school district, as well as their parents and adult volunteers participating in a neighborhood watch program in the district, to carry GPS loggers for 2 weeks. Based on the GPS logger data, information on the location of the children after school was analyzed in terms of whether the adult monitoring activities covered the area. They found that the spatial patterns of the activities of the children changed over time, from routes to and from school to playgrounds, such as parks. However, monitoring activities by the adults primarily focused on the routes, leaving children in playgrounds essentially uncovered.

A common difficulty faced by such neighborhood projects is the availability of GIS tools and individuals capable of handling them. While free GIS software such as QGIS (https://qgis.org/) is increasingly used, effective operation of them could still be demanding for the general public. The Council for the Promotion of Preventive Criminology (CPPC) in Japan provides a simplified GIS tool specializing in basic functionalities to create a map of neighborhood safety based on field surveys (CPPC 2009). Field surveys of a neighborhood can be conducted to identify potential risks by users capturing photographs of items or situations that they think are unsafe with a digital camera and recording comments explaining their concerns with an IC recorder, while simultaneously collecting locational information using a GPS logger. The CPPC tool almost automatically synthesizes the information from the three devices, the digital camera, IC recorder, and GPS logger, according to time stamps and creates a map showing the unsafe locations along with the photographs. This tool is currently available only in Japanese, but similar efforts to help communities utilize GIS and geospatial information are prevalent worldwide.

8.2 Various Applications of Geospatial Information in Crime Prevention

8.2.1 Visualization of Crime Occurrence

As mentioned in Sect. 8.1.1, crime mapping, that is, the visualization of crime occurrence as a map, dates back to the early nineteenth century. It is now commonly employed in conjunction with geospatial information.

In Japan, prefectural police departments have been opening websites that present interactive crime maps, allowing people to visually examine the occurrence of crimes and accidents using Web GIS functionalities. Figure 8.2 shows the crime information map of the Tokyo Metropolitan Police Department (TMPD) as an example of a web-based crime map. The map represents the number of burglary thefts by small address area, a geographic unit similar to census block groups in the U.S., around Shinjuku, Tokyo. Users can also choose other crime categories such as vehicle load theft, motor vehicle theft, shoplifting, and bag lifting. According to the crime statistics provided by TMPD (2021), Shinjuku had the highest number of crimes among the 23 special wards of Tokyo, but the map suggests that only a limited number of the small address areas observed a large number of crimes; the concentration was focused on specific hot spots, such as Shinjuku Station.

Figure 8.3 is another example of a web-based crime map by the Osaka Prefectural Police. This map shows the approximate locations of individual crime incidents by icons indicating crime categories, such as purse snatching. This type of representation allows users to assess in detail where crimes occurred at the street or block level.

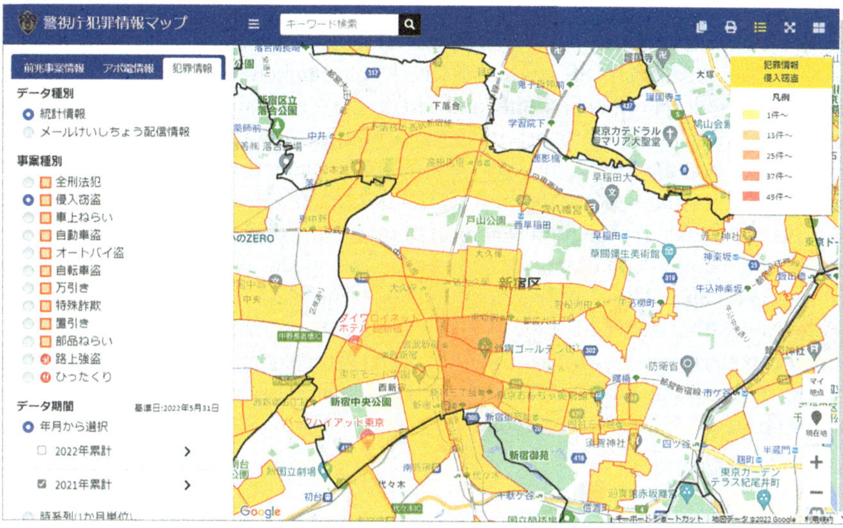

Fig. 8.2 Crime information map of the Tokyo Metropolitan Police Department. (From: https://map.digipolice.jp/)

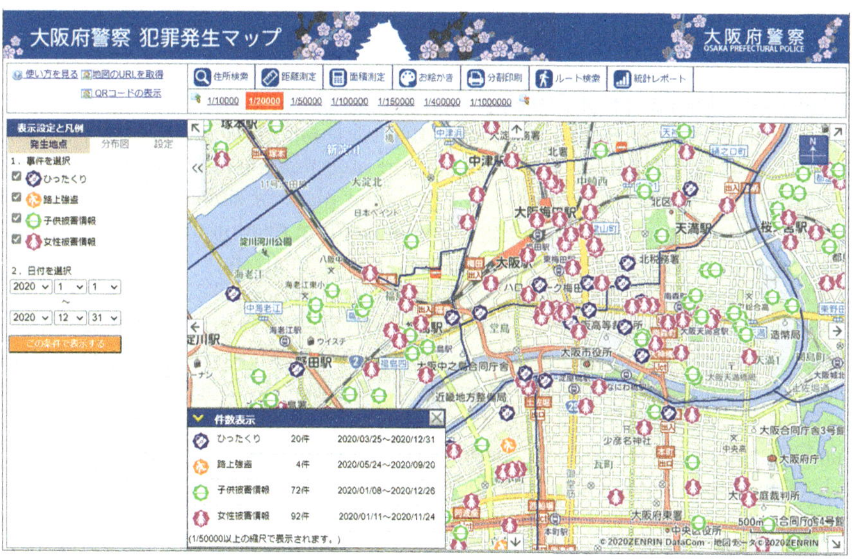

Fig. 8.3 Crime occurrence map of the Osaka Prefectural Police. (From: http://www.machi-info.jp/machikado/police_pref_osaka/index.jsp)

Maps based on aggregate statistics, such as that by the TMPD, are also helpful for capturing the overall spatial distribution of crime. Web-based crime maps are effective tools for communicating areas of high crime risk to the public.

The two examples above use relatively simple visualization techniques to represent the spatial distributions of crime; however, crime maps using more sophisticated methods also exist. Web GIS systems by the Kyoto Prefectural Police (https://kyoto-fukei.geocloud.jp/public/map.html) and Hokkaido Prefectural Police (https://map.police.pref.hokkaido.lg.jp/hp_asp/main.jsp) show crime distributions as density maps. They use the kernel density estimation method (Silverman 1986), which estimates a continuous point density distribution from a discrete point distribution. This method allows for the representation of the spatial distribution of points, in this case, crime locations, that are not affected by the spatial unit of data aggregation. Nakaya and Yano (2008) further extended the kernel density estimation method by incorporating the temporal dimension as the z-axis, so that both the spatial and temporal patterns of a crime distribution can be represented as a 3D map. More advanced spatial statistical methods are also available for identifying crime hot spots more rigorously, but they are beyond the scope of this chapter.

8.2.2 Environmental Factors Related to Crime Risk

Crime research aimed at understanding the mechanisms of crime and delinquency occurrence for prevention or reduction can generally be divided into two categories based on the theory they rely on: crime causation theory and crime opportunity theory. Crime causation theory focuses on the reasons that a person becomes a criminal or delinquent; that is, their personality or environments that tempt them into crime. Crime opportunity theory assumes that someone who can potentially commit a crime exists in all locations and times, and the investigation focuses on why a crime occurred; that is, the vulnerability of a particular place or situation to crime. Both theories consider the environment as a crucial factor in relation to crime occurrence, and thus utilize maps and geospatial information for analysis.

Researchers in the Chicago school mentioned in Sect. 8.1.1 investigated the mechanism of crime occurrence from the perspectives of physical dilapidation and socioeconomic conditions of neighborhoods. Shaw and McKay (1942) analyzed the spatial distribution of juvenile delinquents in Chicago and its associations with the social and demographic characteristics of the neighborhood. Those authors found that high population mobility and poverty were consistently associated with a higher risk of juvenile delinquency, regardless of the racial structure of neighborhoods. They developed the social disorganization theory by assembling findings from various North American cities. The theory states that social order based on traditional customs and norms tends to be disorganized in such disadvantaged neighborhoods, and the social and cultural settings predispose the youth to crime. Crime research at that time was focused on seeking the causes of crime and delinquency based on the hereditary characteristics of the individual. The research by Shaw and McKay opposed the mainstream approach and emphasized the importance of the environments in which people were raised.

GIS, which is highly capable of synthesizing and analyzing information related to specific areas, is well suited to the essential concept of crime causation theory that the environment of the neighborhood induces crime and delinquency. Crime research in this field now broadly utilizes GIS as an important and effective tool.

From the perspective of crime causation theory, crime prevention emphasizes social policies, for example, offering opportunities for education and employment and support for low-income individuals to maintain social order. However, such policy interventions generally require time to achieve notable results. Reflecting on this limitation of crime causation theory, in the 1970s, researchers began to turn their attention to crime opportunity theory, which presumes that the vulnerability of places and situations generates crime opportunities, and to prevent crime, these vulnerabilities must be removed.

Crime opportunity theory is based on the routine activity theory, which states that crime occurs when the following three items come together: a motivated offender, a suitable target, and the absence of a capable guardian (Cohen and Felson 1979). Risks or opportunities for crime exist in our daily routines. For example, a child walking alone after school could be a suitable target of a motivated offender. However, if a neighborhood watch program is implemented there, volunteers on patrol are considered capable guardians who are expected to reduce crime risk. Alternatively, if children walked in a group, they would be less suitable or attractive as targets than a child walking alone. Similarly, houses that are not equipped with security cameras or that cannot be seen from the street would be more attractive for a potential burglar who walks through a neighborhood looking for a target. Capable guardians are not necessarily people intentionally on guard; they also could include equipment such as security cameras and alarm systems, as well as people nearby who could serve as witnesses.

While crime causation theory focuses only on people who could potentially become criminals or delinquents, crime opportunity theory considers the characteristics of places that induce or prevent crime and the spatial behaviors of both potential criminals and potential victims. Further, crime causation theory captures neighborhoods on a relatively large scale, whereas crime opportunity theory tends to focus on smaller areas, such as a corner of a street, a neighborhood playground, or a bicycle parking lot. These characteristics of crime opportunity theory suggest that advances in geospatial information technologies and the increasing availability of detailed geospatial data play a vital role in the expansion of research. The project by Amemiya et al. (2009), introduced in Sect. 8.1.3, is an excellent example of such research benefitting from advanced geospatial information tools, for example, small, lightweight, and low-priced GPS loggers.

Another example comes from Hanaoka (2018) with a study conducted in Osaka City, Japan. This study utilized hourly population data estimated from mobile phone user locations as proxy measures of the spatial distributions of suitable targets and capable guardians. Publicly available population data are generally about residential, namely, nighttime population, but the actual population of an area varies within a day. The population data used here were composed of GPS records of mobile phone users and could capture population changes on small spatial and temporal scales. As

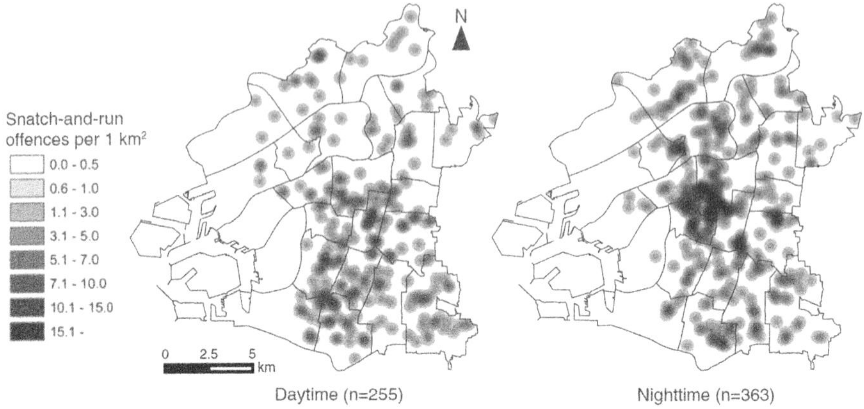

Fig. 8.4 Kernel density maps of snatch-and-run offences in Osaka City. (From: Hanaoka, K. (2018). New insights on relationships between street crimes and ambient population: Use of hourly population data estimated from mobile phone users' locations. Environment and Planning B, 45(2), 295–311)

crime data, Hanaoka used crime alert emails sent by a police department to registered residents. Figure 8.4 presents the daytime and nighttime distributions of snatch-and-run offenses estimated by the kernel density estimation method introduced in Sect. 8.2.1. This study found that associations between populations and snatch-and-run offenses differed between day and night. During the day, offenses were less likely to occur where hourly population density tended to increase, suggesting that the population served as capable guardians. At night, offenses occurred where the population density was relatively high, suggesting that the population represented suitable targets.

8.2.3 Geographic Crime Prediction

It is natural to assume that crime risk is high in areas where crime has been regularly concentrated in the past and areas whose environments are associated with higher crime risk as discussed in the previous section. Crime occurrence also has a tendency called near-repeat victimization, which means that once a crime occurs, similar crimes follow repeatedly within close proximity, in terms of both space and time. This section discusses crime prediction utilizing these geographic characteristics of crime occurrence.

Risk terrain modeling developed by Caplan et al. (2011) estimates the spatial distribution of crime risk by first estimating the densities of facilities that are known or hypothesized to relate to crime risk and then calculating their weighted sum based on the relative importance of the facilities. Risk terrain modeling presents the resulting estimated risk as a continuous surface covering a study region and calls it

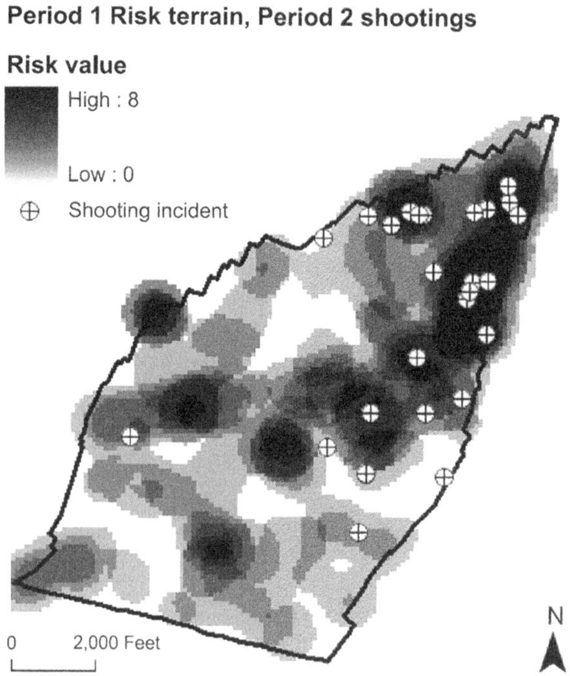

Fig. 8.5 Estimated risk terrain of shootings and locations of actual shootings. (From: Caplan, J. M., Kennedy, L.W. and Miller, J. (2011). Risk terrain modeling: Brokering criminological theory and GIS methods for crime forecasting. Justice Quarterly, 28(2), 360–381)

the "risk terrain" by analogy with the topography. The primary advantages of this method are the straightforward concept of spatially overlaying risk factors, technical simplicity of the computational process that is readily implemented in standard GIS software, and flexibility of the model structure allowing the incorporation of various factors according to the type of crime under study. Note that the latter advantage also means that the choice of factors and their relative weights strongly affects the resulting risk estimation and therefore requires careful consideration.

Caplan et al. (2011) applied risk terrain modeling to forecast the risk or potential of a shooting in Irvington, New Jersey, an inner-city neighborhood where shootings were not uncommon. Those authors selected three key predictors of a shooting based on a literature review: residences of known gangsters; retail infrastructures including but not limited to bars, pawn shops, liquor shops, and fast-food restaurants; and locations of drug arrests. Figure 8.5 shows the estimated risk terrain of a shooting together with the locations of actual shootings, indicating most of the shootings in this study region occurred in areas predicted as high risk by the model.

There are also crime prediction systems that reflect the concept of near-repeat victimization. Such systems utilize spatial and temporal information about past crimes and can generally provide risk estimation on a relatively small scale to support efficient allocation of police resources.

Crime prevention systems have been incorporated into police activity, especially in the U.S., with the expectation that objective data and analysis can support effective policing that is not influenced by human factors. However, there are increasing concerns about crime prediction systems, particularly those based on the concept of near-repeat victimization. Reynolds (2017) suggested a feedback loop in which a new arrest made in an area predicted as high risk becomes an input to the following prediction process, and the same area is again predicted as high risk. This is problematic because the prediction systems cannot differentiate between an increased number of arrests due to intensive patrolling based on the predicted high risk of the area and the actual concentration of crime in that region. Please note that arrests do not necessarily reflect the entire picture of crime. Criticism of crime prevention systems based on the near-repeat victimization concept also points out that the predicted high-risk areas often coincide with neighborhoods with a large proportion of racial minorities, possibly aggravating prejudice and discrimination (Reynolds 2017). Biased police activity poses a serious social problem. The prediction accuracy and crime prevention effects of these systems require careful verification.

8.3 Summary

This chapter discussed various applications of geospatial information in crime prevention efforts after briefly introducing the history of crime analysis using maps and geospatial information. Three major application themes, visualization of crime occurrence, environmental factors related to crime risk, and geographic crime prediction, were discussed in detail, including theoretical bases and recent examples. Studies of subjective perceptions of crime risks and the effectiveness of police activity also utilize geospatial information, although these were not discussed in this chapter. Such applications of geospatial information and techniques in crime prevention are expected to increase and broaden, considering their strong potential to contribute to this field.

References

Amemiya M, Saito T, Kikuchi G, Shimada T, Harada Y (2009) Identifying the spatio-temporal characteristics of children's after school activities and adults' neighborhood watch activities using GPS data. J Jpn Inst Lands Arch 72(5):747–752. (in Japanese)

Bates S (1987) Spatial and temporal analysis of crime. Illin Crim Just Inf Auth Res Bull. https://www.ojp.gov/ncjrs/virtual-library/abstracts/spatial-and-temporal-analysis-crime

Caplan JM, Kennedy LW, Miller J (2011) Risk terrain modeling: brokering criminological theory and GIS methods for crime forecasting. Justice Q 28(2):360–381

Cohen LE, Felson M (1979) Social change and crime rate trends: a routine activity approach. Am Sociol Rev 44:588–608

Council for the Promotion of Preventive Criminology (2009) Prevention of crime against children through science. http://www.skre.jp/nc2/ (in Japanese)

Hanaoka K (2018) New insights on relationships between street crimes and ambient population: use of hourly population data estimated from mobile phone users' locations. Env Plan B 45(2): 295–311

Harada Y (2009) Crime, safety and security and GIS. In: Murayama Y, Shibasaki R (eds) GIS for life and culture. Asakura Publishing Company, Tokyo, pp 97–116. (in Japanese)

Imai O (2015) Introduction and management of GIS in organizations. In: Asami Y, Yano K, Sadahiro Y, Yuda M (eds) Geographic Information Science – GIS Standards. Kokon Shoin Publishers, Tokyo, pp 178–189. (in Japanese)

Nakaya T, Yano K (2008) Spatio-temporal three-dimensional mapping of crime events: visualizing spatio-temporal clusters of snatch-and-run offences. J Geogr 117(2):506–521. (in Japanese)

Reynolds M (2017) A flaw in the pre-crime system. New Sci 236(3146):10

Shaw CR, McKay HD (1942) Juvenile delinquency in urban areas. University of Chicago Press, Chicago

Silverman BW (1986) Density estimation for statistics and data analysis. Chapman & Hall/CRC, Boca Raton, FL

Chapter 9
Applications in Health and Healthcare

Ikuho Yamada

Abstract Health and healthcare are one of the research fields where applications of geospatial information are expected to enhance analyses of related issues and support decision-making processes. Applications in this field cover a wide variety of users and range from quantitative analysis of the spatial distribution of a particular disease to web-based search services of healthcare facilities for the general public. This chapter introduces the applications of geospatial information in the field of health and healthcare, focusing on the spatial distributions of diseases, locations of healthcare facilities and services, and associations between human health and the environments.

Keywords Disease maps · Spatial diffusion of disease · Spatial surveillance · Demand prediction for healthcare facilities and services · Accessibility to healthcare facilities and services · Human health and environments

9.1 Spatial Distribution of Diseases

9.1.1 Disease Maps

Mapping the occurrence of disease cases is a simple but widely used application of geospatial information in the field of health and healthcare. The first step in detecting and understanding the characteristics of the spatial distribution of a phenomenon occurring on the Earth's surface is to represent the phenomenon as a map and examine it visually. The field of health and healthcare is no exception. Maps of the number of disease cases per small zone or its ratio to the risk population help us recognize the regional disease risk disparity, disease hot spots where an abnormally large number of cases are concentrated, or other factors. The detection of such

I. Yamada (✉)
Center for Spatial Information Science, The University of Tokyo, Chiba, Japan
e-mail: iku.yamada@csis.u-tokyo.ac.jp

Y. Sekimoto, Y. Kawahara (eds.), *Utilization of Geospatial Information in Daily Life*, New Frontiers in Regional Science: Asian Perspectives 65,
https://doi.org/10.1007/978-981-19-6905-8_9

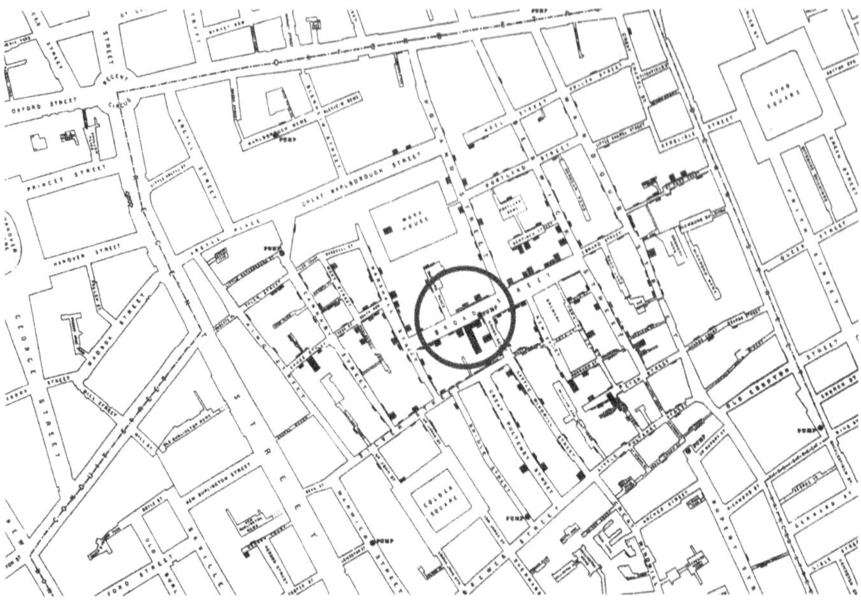

Fig. 9.1 Snow's cholera map. (From: Snow, J. (1855). On the Mode of Communication of Cholera (2nd ed.). London: John Churchill. https://commons.wikimedia.org/wiki/File:Snow-cholera-map-1.jpg. Part of the original map was extracted, and a circle at the center was added by the author)

high-risk areas allows the effective allocation of resources for disease control measures. Further, by combining other geospatial information, such as the demographic and climate characteristics of regions, we can proceed to investigate whether and how regional characteristics relate to disease risk. Such maps that present information about diseases and health are called disease maps.

Disease maps came to be used in epidemiological research in Western countries at the end of the eighteenth century. One of the most famous disease maps is the cholera map created by Dr. John Snow, an English physician, when a cholera outbreak occurred in London in 1854 (Fig. 9.1). Based on his experience as a physician, Snow noticed that the spatial distribution of cholera cases was related to the service areas of different water companies. He pointed out that a water pump located in the center of the concentration of cases was the source of infection and brought an end to the outbreak.

A map presenting the locations of individual cases as points is called a dot map. The cholera map described above is a variant. Other types of maps commonly used for disease mapping are the choropleth map and the proportional symbol map shown in Fig. 9.2. Both display quantitative attributes of the zones. In the choropleth map, each zone is shaded according to the value of the attribute of interest. In the proportional symbol map, each zone is represented by a geometric feature, such as a circle and a square, whose size is proportional to the attribute value. The former is appropriate for quantitative attributes that consider the size or population of each

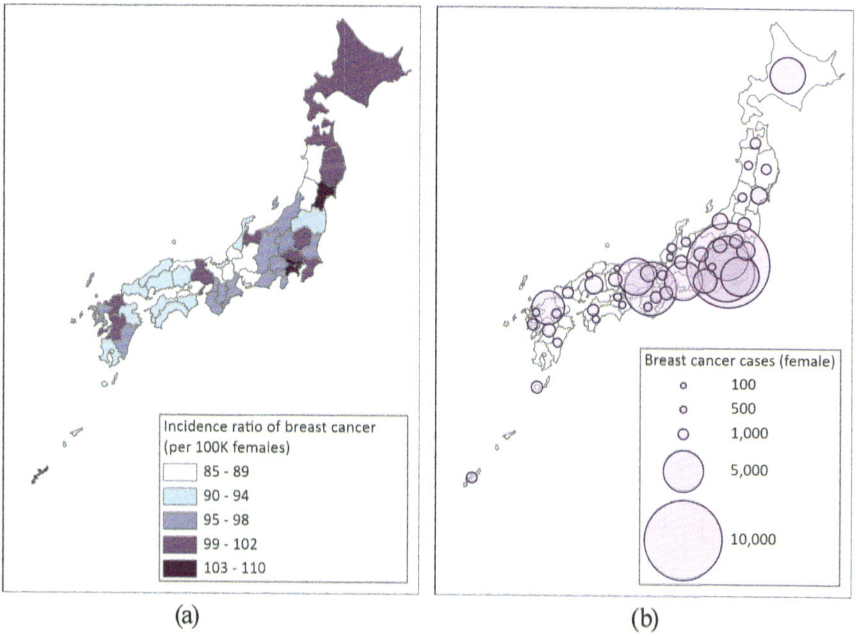

Fig. 9.2 Disease maps representing the occurrence of breast cancer in Japan. (**a**) Choropleth map, (**b**) Proportional symbol map (Data source: National Cancer Registry in Japan 2018)

zone, for example, disease rate and population density. The latter is used for quantitative attributes representing total counts or amounts.

The progress of geospatial information techniques has made more advanced, sophisticated map representations possible. Figure 9.3 shows an example of such an advanced map. Both maps in Fig. 9.3 are 3D representations of the distribution of the standardized mortality ratio (SMR) by municipality in Japan, where the shading and height of each municipality are determined according to its SMR. The upper map is a 3D version of the standard choropleth map. In contrast, the lower map is based on a cartogram where the size of each municipality is scaled in relation to its population, and the volume of each municipality is determined to represent its excess death cases. Comparing the two maps, we can see that the local concentrations of death cases and regional inequality in the risk of death are more clearly illustrated in the lower map.

The usefulness of geospatial information in the field of health and healthcare has been recognized by internationally known health-related organizations such as the World Health Organization (WHO) and the Center for Disease Control and Prevention (CDC) in the U.S. These organizations contribute to the promotion of geospatial information in the field by not only opening geospatial information on diseases and public health to the public but also by providing GIS software specializing in functionalities needed in the field. For instance, the CDC offers free software for epidemiological analysis called Epi Info (https://www.cdc.gov/epiinfo/index.html).

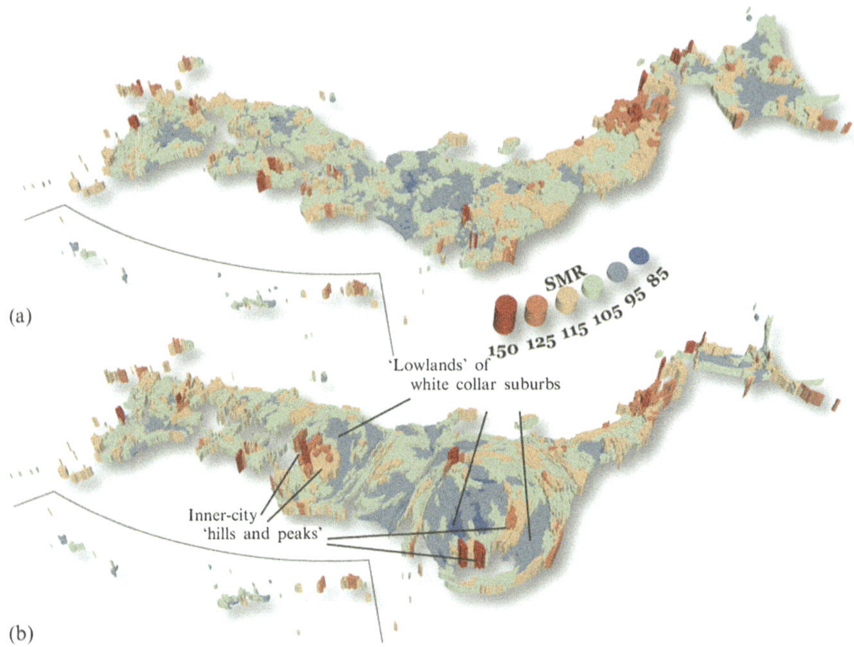

Fig. 9.3 SMR distribution in Japan; (**a**) 3D choropleth map; (**b**) 3D cartogram. (From: Nakaya, T. (2010). 'Geomorphology' of population health in Japan: looking through the cartogram lens. Environment and Planning A, 42(12), 2807–2808)

This software provides functionalities to create dot and choropleth maps and helps public health practitioners and researchers easily synthesize epidemiologic analyses with geospatial information techniques.

In addition to visualizing the spatial distributions of disease and health risks, geospatial information also contributes to the quantitative, objective analysis of health-related issues utilizing methodologies in statistics and mathematical modeling. Later sections introduce examples of such applications.

9.1.2 Spatial Diffusion of Disease

The process through which a disease gradually spreads from a region where it first occurs to the surrounding regions is called the spatial diffusion of disease. While a disease map generally deals with the spatial distribution of disease at a particular point in time, spatial diffusion considers spatio-temporal changes in disease distribution. The concept of spatial diffusion is broadly used to model a mechanism through which phenomena such as new knowledge, innovation, and rumors spread over regions. In the field of health and healthcare, it is often applied to the spread of infectious diseases, as well as new medical treatments and medicines.

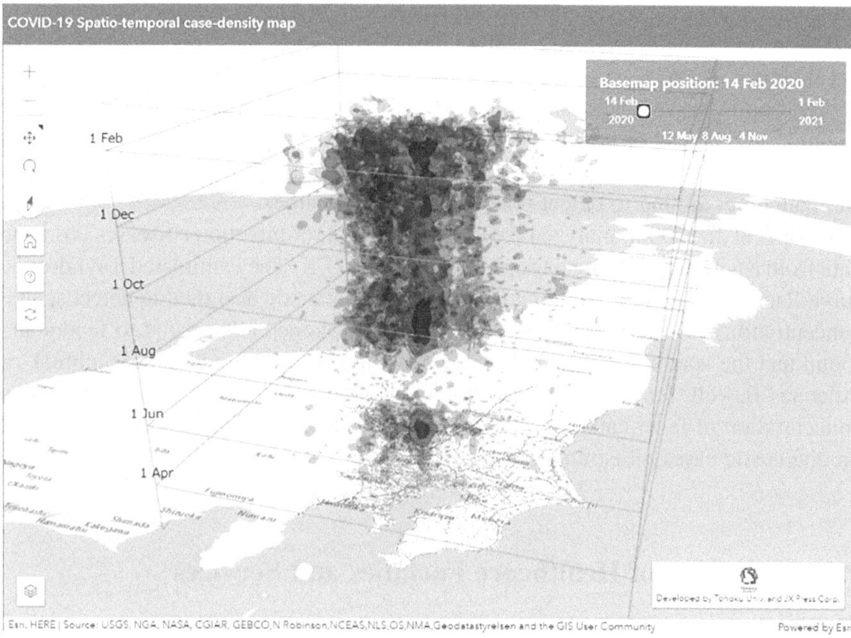

Fig. 9.4 COVID-19 spatio-temporal case-density map developed by the Graduate School of Environmental Studies, Tohoku University, and JX Press Corporation. (From: https://nakaya-geolab.com/covid19-stkd/tokyo/)

Visualizing how the disease has spread over space on a map itself helps explore the mechanism of the diffusion process in the same way as disease maps. Currently, more advanced analyses using detailed geospatial data and quantitative analytical methodologies are gaining popularity. Figure 9.4 shows a 3D map representing the spatial diffusion of COVID-19 around the Tokyo metropolitan area, Japan, developed by the Graduate School of Environmental Studies, Tohoku University, and JX Press Corporation. In this map, the horizontal axes show the geographic space and the vertical axis shows the time. The cloud-like shapes represent areas with a certain level of a high infection density, which was estimated based on the locations of facilities where outbreaks had occurred over a period of time.

9.1.3 *Spatial Surveillance of Disease*

Spatial analysis can be divided into *retrospective* and *prospective* studies in terms of the temporal scale. The methods discussed in this chapter so far are retrospective in the sense that they look back at the spatial pattern of a disease distribution or its changes based on data observed at a point (or points for spatial diffusion) in the past. Spatial surveillance, on the other hand, is a prospective analysis that examines time-series data of the spatial distribution consecutively as recent data become available to detect changes in the distribution as soon as possible. The increasing availability of real-time and nearly real-time spatial data for various phenomena supports the development of this type of analysis.

Syndromic surveillance, also known as biosurveillance, has recently attracted attention in the field of health. While disease surveillance generally monitors cases diagnosed by doctors, syndromic surveillance examines symptoms, such as fever and coughing, aiming to detect potential outbreaks earlier.

Cooper et al. (2008) analyzed the spatial diffusion of infectious diseases based on data from a national telephone triage service in the U.K. They simulated a syndromic surveillance situation by applying the spatial scan statistic designed to detect spatial concentrations of cases (Kulldorff et al. 2005) to phone calls related to fevers and found that the spatio-temporal pattern of the calls reflected the national outbreak of influenza B well. Further, the syndromic surveillance detected a significant spatial concentration of fever calls much earlier than the influenza outbreak detected based on diagnostic cases, illustrating its effectiveness.

9.2 Locations of Healthcare Facilities and Services

9.2.1 Demand Prediction for Healthcare Facilities and Services

Geospatial information is also helpful in capturing the demands of healthcare facilities and services, as well as in planning and evaluating their locations. Identifying areas with high health risks, for example, areas where the older adults or infants are concentrated, allows for the effective allocation of healthcare resources and efficient provision of healthcare services. Objectively assessing whether the current locations of healthcare facilities ensure reasonable accessibility for residents is an essential measure of spatial decision support for adding new facilities or reallocating existing ones.

McLafferty and Grady (2004) predicted the demand for prenatal care clinics using data on the residential locations of recent mothers in need and compared it with the distribution of existing prenatal care clinics. The comparison was based on density maps generated using the kernel density estimation method introduced in Chap. 8. By combining the two density maps, they identified high-priority areas for new clinics that observed both a high density of mothers in need and a low density of existing clinics nearby.

While McLafferty and Grady's study focused on a single borough in New York, there are also recent studies handling a much larger region. Doi et al. (2015) estimated the supply and demand balance of medical services at a spatial resolution of 500 × 500 m for the Tokyo metropolitan area, consisting of Tokyo and its three neighboring prefectures. The estimation was based on a simulation of patients' use of medical institutions, taking into account future estimates of the number of inpatients and travel time to the institutions. They examined temporal changes in the excess demand for medical services from 2010 to 2040, with a specific focus on the spatial distribution of areas facing a shortage of inpatient beds in hospitals.

9.2.2 Accessibility to Healthcare Facilities and Services

The concept of accessibility refers to how easily people can access facilities, which can be affected by the spatial configurations of people and facilities, as well as by their socioeconomic or cultural situations. Geospatial information and GIS are broadly utilized to assess the physical and spatial aspects of accessibility in such research fields as healthcare, public transportation, marketing, and education.

The most straightforward measure of spatial accessibility is the distance to the nearest facility, that is, a hospital. When a study region has areas where multiple hospitals are concentrated, a gravity model capturing the spatial interaction between locations over distance would be more suitable to reflect the advantage of having access to multiple hospitals. The 2-step floating catchment area method (2SFCA) developed by Luo (2004) and its enhanced variants are now considered a standard method for the assessment of spatial accessibility, incorporating not only the distance to the medical facilities but also the service availability relative to the population size. Please note that "distance" here covers a broader concept than the straight-line distance, including the shortest path distance along the road network and travel time reflecting temporal variations in travel speed, which can be obtained using geospatial information.

Langford et al. (2016) applied the 2SFCA method to assess accessibility to primary healthcare in South Wales, U.K. They enhanced the analysis by differentiating public and private transportation modes, namely cars and buses, using network datasets developed dedicatedly for each mode. Figure 9.5 shows the spatial distributions of the accessibility to primary healthcare as the number of general practitioners available per 1000 population by walking and buses (map (a)) and by driving (map (b)). These maps imply inequalities in accessibility to healthcare, not only in terms of individuals' residential locations but also in terms of transportation modes available to them.

9.3 Human Health and Environments

9.3.1 Potential Associations Between Human Health and Environments

Except for certain hereditary diseases, the occurrence of any other disease is associated with external environmental factors. A research field in geography that

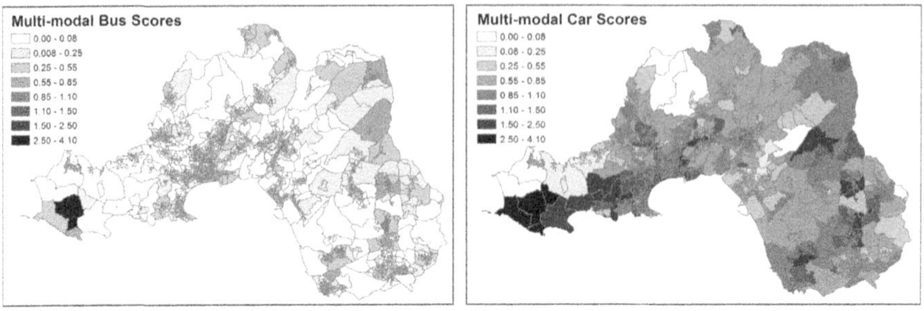

(a) (b)

Fig. 9.5 Accessibility to primary healthcare by public and private transportation modes. (**a**) public transportation only; (**b**) private transportation. (From: Langford, M., Higgs, G., and Fry, R. (2016). Multi-modal two-step floating catchment area analysis of primary health care accessibility. Health & Place, 38, 70–81)

concerns such associations is called medical geography, whose origin is said to be traced back to Hippocrates, a Greek physician known as the "Father of Medicine." In his book *On Airs, Waters, and Places*, Hippocrates explained the importance of considering environmental factors such as season, wind, climate, water quality, and sunshine to properly investigate medicine. As geospatial information has an outstanding capability of handling information about the environment, it has an important role in the investigation of environmental factors that promote or hinder human health.

Traditionally, medical geography mainly examines physical environments, as they are closely associated with infectious and nutrition-related diseases, which have long been at the center of health problems. Now that the focus has shifted to lifestyle-related illnesses and chronic diseases, risk factors for human health have come to include much broader and more complex elements. The new, more comprehensive concept of "environments" contains the physical structures of cities, locations of healthcare facilities, social systems, and so on, in addition to physical environmental factors. Similarly, the research themes that medical geography deals with have also expanded, and it is now commonly referred to as health geography. The following two sections discuss the physical environments and broader concept of living environments.

9.3.2 Physical Environments and Health

Similar to disease maps, mapping an environmental factor is helpful for understanding and exploring the characteristics of its spatial distribution and potential associations with disease distribution. Geospatial information on physical environments, such as water and air quality, has gradually become available to the public through government agencies or municipal offices. In the U.S., the Environmental Dataset

Gateway (https://edg.epa.gov/) managed by the Environmental Protection Agency (EPA) and GeoPlatform.gov (https://www.geoplatform.gov/) managed by the Federal Geographic Data Committee provide free geospatial information on physical environments. The National Institute for Environmental Studies in Japan offers a web-based GIS tool that allows the interactive mapping of environmental factors (http://tenbou.nies.go.jp/).

In the case of infectious diseases, the physical environments influence their occurrence via the spatial distribution of viruses and vectors. For example, the risk of malaria infection is strongly related to the distribution of mosquitoes, which transmit causative parasites to people. Although it is possible to survey their habitat using mosquito traps, it is too costly to cover a large study region. Researchers have thus come to use geospatial information to estimate it by developing models that explain the potential of mosquitoes' living based on data on physical environments such as vegetation and temperature. The estimated habitat of mosquitoes is further combined with geospatial demographics data to model a disease epidemic efficiently and effectively.

Bouzid et al. (2014) developed a model to predict future epidemics of dengue fever in the EU. They used maximum and minimum temperatures, precipitation, and humidity as climatic variables, and population density, population size, urbanization, and GDP as sociodemographic variables. Using a model calibrated with data on Mexico around the end of the twentieth century, they predicted the number of dengue fever cases per 10 × 10 km grid cell during the twenty-first century in the EU under three climate change scenarios. Figure 9.6 shows an example of the predicted situations.

9.3.3 Broader Concept of the Living Environments and Health

One reason why the broader concept of living environments has attracted increasing attention is the growing epidemic of obesity in developed countries, such as the U.S. and the U.K. While obesity has long been considered a personal problem caused by individuals' lifestyles, research and public health countermeasures based on that idea have so far failed to improve the obesity problem at the population level. This situation has fostered the idea that environments beyond the control of individuals induce obesity, and elements of living environments that promote or prevent obesity require close investigation. Living environments may be related to obesity through physical activity and eating habits, and this section introduces an example of the former, *walkability*.

Walkability refers to environmental support for walking, which is expected to encourage people to walk instead of driving, and therefore increase their physical activity level. Walkable neighborhoods are often characterized by indices called *5Ds*: population density (*D*ensity), land-use diversity (*D*iversity), pedestrian-

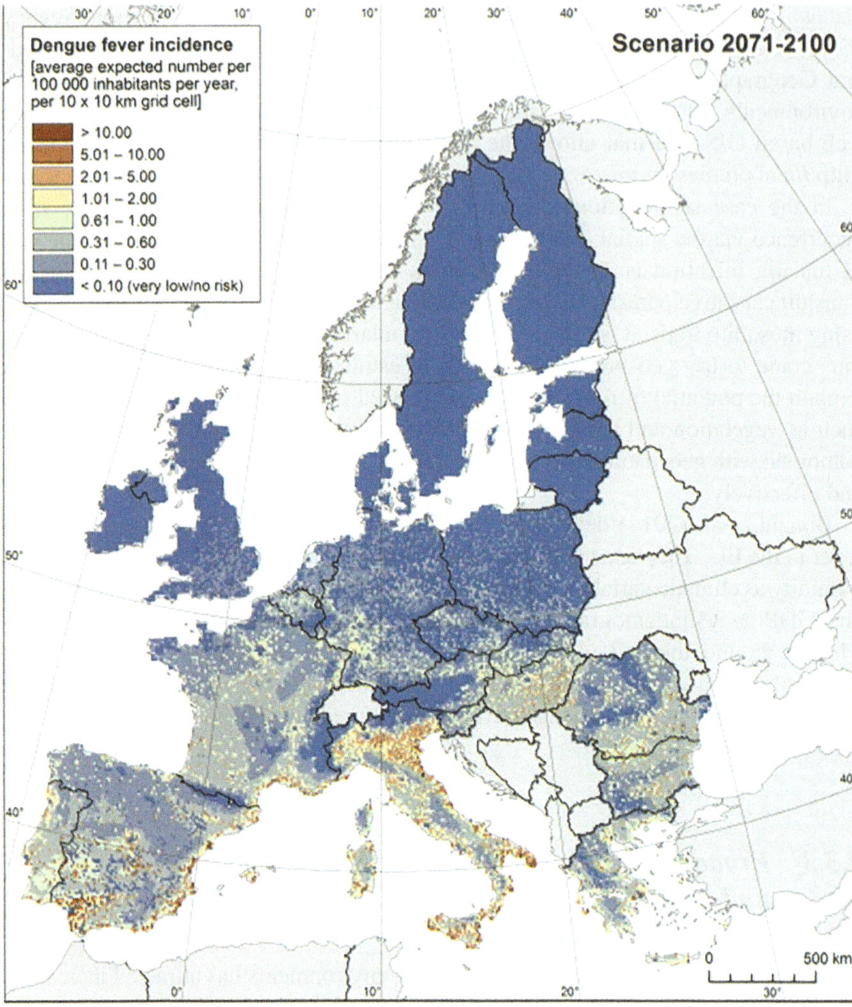

Fig. 9.6 Predicted incidence rate of dengue fever in the EU under the climate change scenario for the late century. (From: Bouzid, M., Colón-González, F.J., Lung, T., Lake, I.R., and Hunter P.R. (2014). Climate change and the emergence of vector-borne diseases in Europe: case study of dengue fever. BMC Public Health 14, 781. https://doi.org/10.1186/1471-2458-14-781)

friendly design (*D*esign), destination accessibility (*D*estination), and distance to transit (*D*istance) (Cervero and Kockelman 1997; Ewing and Cervero 2001). It is hypothesized that the better each index is, the more people will walk in their daily lives and live healthier lives.

Data on these walkability indices may be collected through questionnaires and field surveys, but geospatial information on neighborhood built environments enables easy computation or estimation for a larger study region. Health geography research compares the spatial distributions of the indices with those of residents' health indicators, such as obesity rate and average number of steps walked per day, using quantitative analytical methods to identify potential relationships between them.

A study in Salt Lake City, Utah, in the U.S. (Yamada et al. 2012) defined the area of an individual's daily living life as an area that the individual could reach by walking 1 km along the streets from their residence, computed with network analysis functionalities of GIS. This study examined various walkability indices for the area of daily living, such as population density, intersection density, land-use mixture, and accessibility to urban facilities, in terms of their association with residents' body mass index (BMI). Walkability research originating in the obesity problem of Western countries is currently carried out worldwide, and the properties of walkable, healthy neighborhoods have gradually been revealed.

9.4 Summary

This chapter discussed the basic concepts and methodologies related to the application of geospatial information in the field of health and healthcare. The three major applications covered are the spatial distributions of diseases, locations of healthcare facilities and services, and human health and environments. The examples of applications introduced in this chapter suggest the great potential of geospatial information and techniques to contribute to this field.

References

Bouzid M, Colón-González FJ, Lung T, Lake IR, Hunter PR (2014) Climate change and the emergence of vector-borne diseases in Europe: case study of dengue fever. BMC Public Health 14:781. https://doi.org/10.1186/1471-2458-14-781

Cervero R, Kockelman K (1997) Travel demand and the 3Ds: density, diversity, and design. Trans Res D-Trans Env 2(3):199–219

Cooper DL, Smith GE, Regan M, Large S, Groenewegen PP (2008) Tracking the spatial diffusion of influenza and norovirus using telehealth data: a spatiotemporal analysis of syndromic data. BMC Med 6:16

Doi S, Ide H, Inoue T, Katayama Y, Nishide A, Nakamura T, Fujita S, Suzuki T, Takabayashi K (2015) Estimation and evaluation of health services needs and demand in Tokyo metropolitan area based on a patient access area model. Jpn J Med Inf 35(4):157–166. (in Japanese)

Ewing R, Cervero R (2001) Travel and the built environment. Transp Res Rec 1780:87–114

Kulldorff M, Heffernan R, Hartman J, Assunção RM, Mostashari F (2005) A space-time permutation scan statistic for the early detection of disease outbreaks. PLoS Med 2:216–224

Langford M, Higgs G, Fry R (2016) Multi-modal two-step floating catchment area analysis of primary health care accessibility. Health Place 38:70–81

Luo W (2004) Using a GIS-based floating catchment method to assess areas with shortage of physicians. Health Place 10(1):1–11

McLafferty S, Grady S (2004) Prenatal care need and access: a GIS analysis. J Med Syst 28(3): 321–333

Nakaya T (2010) 'Geomorphology' of population health in Japan: looking through the cartogram lens. Environ Plan A 42(12):2807–2808

Yamada I, Brown BB, Smith KR, Zick CD, Kowaleski-Jones L, Fan JX (2012) Mixed land use and obesity: an empirical comparison of alternative land use measures and geographic scales. Prof Geogr 64(2):157–177

Chapter 10
Application in Times of Disaster: Through the Great Hanshin Earthquake, the Great East Japan Earthquake, and COVID-19

Yoshihide Sekimoto

Abstract Japan is a country where various disasters occur, such as earthquakes, tsunamis, floods, slope collapses, heavy snowfalls, and volcanic eruptions. We will take a bird's eye view of geospatial information from the aspect of information technology that comprehensively responds to these disasters, including the history of the Great Hanshin Earthquake and the Great East Japan Earthquake, as well as the recent response to COVID-19.

Keywords The Great Hanshin Earthquake · The Great East Japan Earthquake · COVID-19 · Government trend · Mobile communication information · Disaster simulation · Disaster archive

10.1 Lessons from the Great Hanshin Earthquake

10.1.1 Geographic Information Utilization in Disasters Has Begun in Earnest

In this chapter, I would like to discuss how geographic information has been used in times of disaster and how it will be used in the future. While the use of geographic information in the case of man-made disasters was for the management of underground facilities as described in Chap. 6, Japan is a country where various large-scale natural disasters such as earthquakes, tsunamis, floods, slope collapses, heavy snowfalls, and volcanic eruptions can occur. It was not until the Great Hanshin Earthquake in January 1995 that the need to prepare for such disasters was recognized. It is said that there were about 640,000 houses that were partially damaged or more (Hyogo Prefecture 2008), and this brought to the fore the need to analyze the earthquake resistance and fire spread of buildings, and to issue disaster prevention

Y. Sekimoto (✉)
Center for Spatial Information Science, University of Tokyo, Tokyo, Japan
e-mail: sekimoto@iis.u-tokyo.ac.jp

Y. Sekimoto, Y. Kawahara (eds.), *Utilization of Geospatial Information in Daily Life*, New Frontiers in Regional Science: Asian Perspectives 65,
https://doi.org/10.1007/978-981-19-6905-8_10

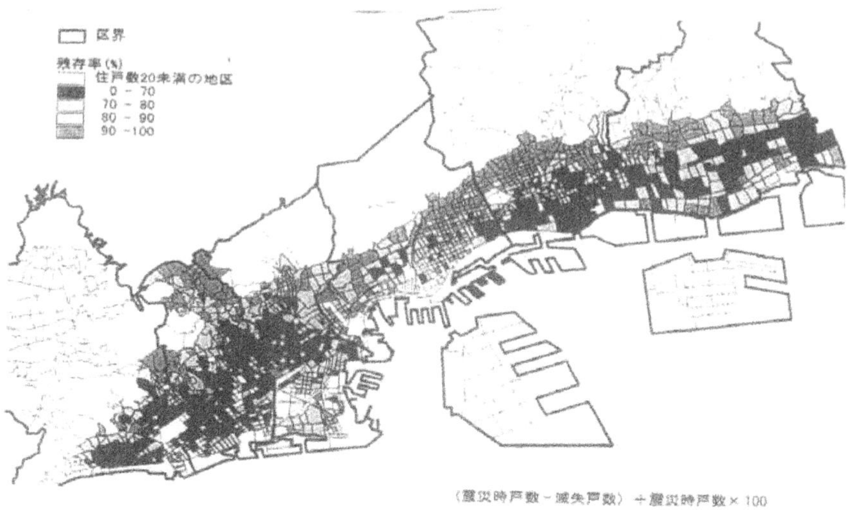

Fig. 10.1 Remaining housing conditions after the Great Hanshin Earthquake. Darkest areas are under 70% for remaining buildings. (Cited from Kobe City Report, Kobe City 2011)

certificates accordingly. In order to issue disaster certification, the authorities conducted damage surveys for each building and created a GIS to understand and visualize the status of remaining buildings in each area (Fig. 10.1), while research institutes conducted studies on the relationship with faults and the spread of fire (for example, the Building Research Institute 1996; Murao 1999).

10.1.2 Government System in Place

Along with the above analysis, the development of a national geographic information (GIS) system also began at this time. In September 1995, after the Great Hanshin Earthquake struck, a liaison conference of ministries and agencies related to geographic information systems (GIS) was established, and a long-term plan for the development of national spatial data infrastructure and the promotion of GIS was decided the following year, followed by an action program in 2002. The following year, a long-term plan for the development of a national spatial data infrastructure and promotion of GIS was established, followed by an action program in 2002. Later, expectations for the launch of the Quasi-Zenith Satellite, a domestically produced GPS satellite, were raised, leading to the establishment of the Basic Act on the Promotion of Utilization of Geospatial Information in 2007. Later, the Council for the Promotion of Geospatial Information Utilization was established in 2008, and in response to the Great East Japan Earthquake, the Cabinet approved the new Basic Plan for the Promotion of Geospatial Information Utilization in March 2012 (Table 10.1) (Cabinet Office 2011).

Table 10.1 History of Japanese government policies starting from the Great Hanshin Earthquake

Year	Action
1995.9	Establishment of the liaison conference of ministries and agencies related to geographic information systems (GIS)
1996.12	Decision of "long-term plan for the development of National Spatial Data Infrastructure and promotion of GIS"
1999.3	Decision of "national spatial data infrastructure standards and development plans"
2002.2	Decision of "GIS Action Program 2002–2005"
2005.3	Establishment of "joint subcommittee on positioning and geographic information systems" by the Liberal democratic party. In September of the same year, the cabinet established the "Council for the Promotion of positioning and geographic information systems"
2006.5	The basic bill for the promotion of geospatial information utilization was submitted to the diet by the Liberal Democratic Party and others
2007.5	The law is passed by the plenary session of the house of representatives, and is enacted and promulgated
2008.4	The basic plan was approved by the cabinet
2012.3	The new "basic plan for the promotion of utilization of geospatial information approved by the cabinet. In October of the same year, an action plan (G-spatial action plan) was formulated

10.2 The Great East Japan Earthquake

10.2.1 Emergency Phase

Sixteen years after the Great Hanshin Earthquake, the Great East Japan Earthquake of March 11, 2011, is still fresh in our minds. To put it simply, the major changes that have taken place are, for better or worse, the ease of information dissemination and the increase in the volume of information from the general public, following the so-called "Web 2.0" period in the mid-2000s, when search engines, SNS, blogs, etc. became widely popular. However, in my personal opinion, especially in times of emergency, such as large-scale disasters, when people are forced to rely on their own judgment, it would be better if information is transmitted from many sources in a grassroots manner. I am not talking only about geographic information.

Although not limited to geographic information, Twitter, with its short and limited number of words, and its simplicity that makes it easy to publish messages, was used during the earthquake and was widely covered (Ministry of Internal Affairs and Communications 2011). It may have been most suitable for such emergencies. For example, Fig. 10.2 shows a mapping of Twitter in the Tokyo metropolitan area at the time of the earthquake, from keywords to locations.

However, compared to the Tokyo metropolitan area, where there was little physical damage, the Tohoku region suffered so much damage from the tsunami that IT was powerless to help. It was reported that people's behavior at that time included "pick-up behavior," in which people who went to help were swept away by

Fig. 10.2 Twitter tweets in the Tokyo metropolitan area at the time of the Great East Japan Earthquake. (Cited from: http://media.mapping.jp/, *Not currently available)

the tsunami, and the "deadlock phenomenon," in which entire streets became stuck in traffic jams due to vehicle congestion (Abe 2014).

10.2.2 Recovery Phase

After that, in the recovery phase, various kinds of support using geographic information were provided. These are sometimes referred to as Emergency Mapping, and the first one to be mentioned is "Sinsai.info (http://sinsai.info)." The Open Street Map Foundation Japan, which supports the Japanese branch of the global OpenStreetMap volunteer-based map, took the lead in setting up this site within 3 h of the earthquake (Fig. 10.3) (Furuhashi 2011). Using Ushahidi, a simple open-source Web GIS software, anyone can post information on lifelines, evacuation centers, safety confirmation, etc., which is the first thing needed after a disaster. The information can be uploaded in real time via Twitter, e-mail, form submissions, and other information sites. We also made efforts to disseminate information to foreign countries, and actively translated information into many languages. This site is a groundbreaking example of how crowdsourcing technology, which allows not only a few experts but also the general public to bring together information they know, can function properly in an emergency situation. The details of these systems such as OpenStreetMap are described in Chap. 14.

In addition to Sinsai.info, other sites that were in operation immediately after the earthquake include the Great East Japan Earthquake Collaborative Information Platform (All311: http://all311.ecom-plat.jp/) by the National Research Institute for Earth Science and Disaster Prevention and the Emergency Mapping Team

Fig. 10.3 Sinsai.info. (Cited from http://sinsai.info. *Not currently available)

(EMT: http://www.drs.dpri.kyoto-u.ac.jp/emt/) by the Disaster Prevention Research Institute of Kyoto University.

In addition, immediately after the earthquake, there were many roads that were closed to traffic, but it was not easy to get a complete picture of these roads. The GPS information installed in the car navigation systems of each manufacturer is aggregated, and since each manufacturer has several hundred thousand to one million units of GPS information, by overlaying all the information, we can get objective data on whether the roads were passable or not. In fact, 8 days after the earthquake, ITS Japan, a nonprofit organization, released a map of automobile traffic records (http://www.its-jp.org/saigai/), which is updated daily, and the Geospatial Information Authority of Japan (GSI) provided a map of road closure information (Fig. 10.4). We have heard from some logistics companies that this has made it easier to provide logistics support to the affected areas.

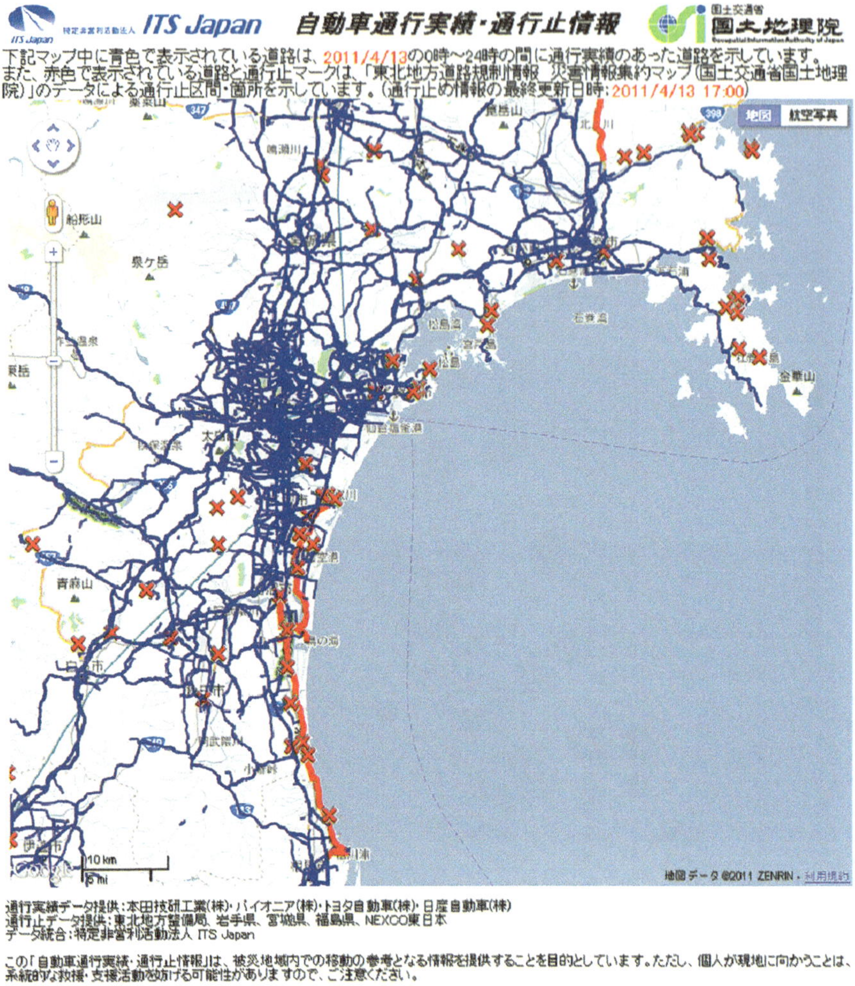

Fig. 10.4 ITS Japan's "Vehicle and Traffic Performance Information Map." (Cited from http://www.its-jp.org/saigai/)

There was also support from the air. Domestically, many organizations, mainly the Geospatial Information Authority of Japan (GSI), provided aerial photographs based on disaster agreements. In addition, as a global framework, there is the International Charter for Disaster (https://www.disasterscharter.org/), which is an international organization that provides and analyzes satellite remote sensing images. Immediately after the earthquake, JAXA made a request for images to be taken, and the participating organizations took images and analyzed them when Japan was in the area. Since this is a global framework for disasters around the world, it is an important initiative in terms of providing rapid support when resources are limited in the immediate aftermath of a major disaster (these details are discussed in Chap. 13).

10.2.3 Preservation of Long-Term Memories

Finally, I would like to talk about how to preserve the lessons of the earthquake for future generations from the perspective of long-term preservation. In this sense, several major archives have been established, including Tohoku University's "Michinoku Shinrokuden" (http://shinrokuden.irides.tohoku.ac.jp/) and the National Diet Library's "Hinagiku" archive (http://kn.ndl.go.jp/). The significance of this is great. In this section, I would like to introduce the Earthquake Disaster Reconstruction Support Research Archive, which I was involved in as a member, in order to promote open government and open data, which have been one of the pillars of IT policy in Japan since 2012.

I would like to introduce the Earthquake Disaster Reconstruction Support Research Archive, which I was involved in as a member of the project. Establishing and disseminating a system for storing and utilizing these records in a centralized manner that can be accessed by anyone is one of the most important things. Based on these policies, the Urban Planning Bureau of the Ministry of Land, Infrastructure, Transport and Tourism has conducted a comprehensive and systematic survey of the affected municipalities in the form of the Great East Japan Earthquake Tsunami Disaster Urban Area Reconstruction Support Survey (hereinafter referred to as the "Reconstruction Support Survey").

For details, please refer to Sekimoto et al. (2013). The authors constructed the "Reconstruction Assistance Survey Archive (http://fukkou.csis.u-tokyo.ac.jp)" to handle these surveys. Specific data items include tsunami inundation, damage to buildings, evacuation methods for individuals and businesses, victims (dead and missing), damage to public facilities and lifelines, and damage to educational facilities and cultural assets. In terms of the quantity of files, there are a total of about 90 types, 114,000 files, and over 200 GB in size.

These valuable data sets have revealed some new things to us. Figure 10.5 shows an animation of the ever-changing evacuation situation of 501 people in Rikuzentakata City, based on shape data of evacuation routes of "Evacuation method (individual)" from the interview survey. Since the start and end time of the evacuation are included in the attributes of the shape data, the approximate location of people at each time can be determined by spatio-temporal interpolation. In addition, the "evacuation zone" data is superimposed on the background (gray area).

As a result, we can see that before the earthquake, people were active as usual (Fig. 10.5a); 15 min after the earthquake, many people started to evacuate, but the situation was quite confusing (Fig. 10.5b); just before 15:25, when the tsunami was said to have hit Rikuzentakata City, there were still people to some extent (Fig. 10.5c); and at around 16:00, there were people in the areas adjacent to the inundation zone and in the (Fig. 10.5d), and at around 16:00, people evacuated to

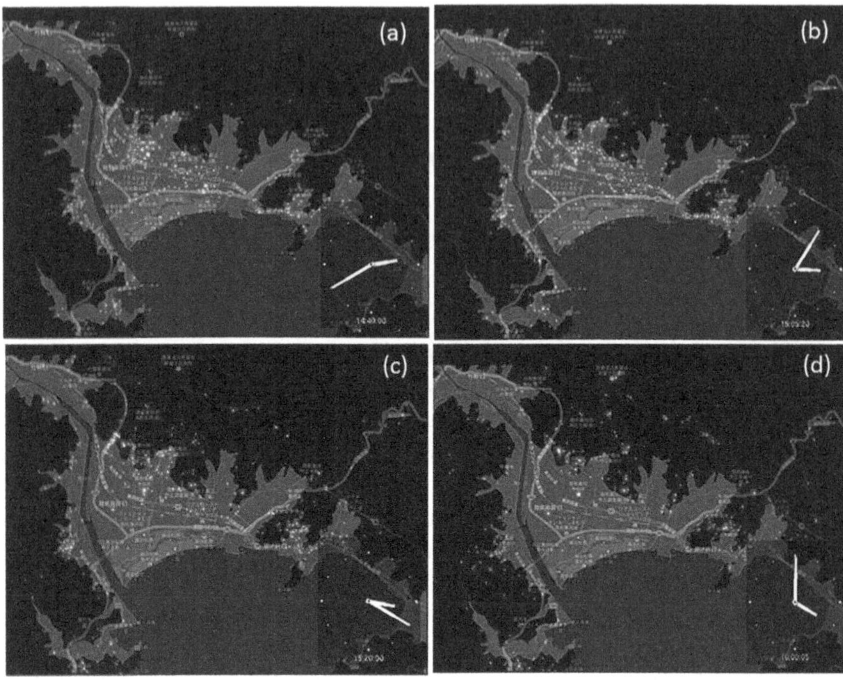

Fig. 10.5 Animation of the evacuation situation in Rikuzentakata City on the day of the earthquake based on the "evacuation route (individual)" interview survey in the earthquake reconstruction support archive. (Sekimoto et al. (2013). (Gray is "inundation area." (**a**) shows the situation of daily activities before the earthquake, (**b**) shows the situation of trying to evacuate to various places immediately after the earthquake, (**c**) shows the situation just before the tsunami, and (**d**) shows the situation of evacuation to outside the evacuation zone or to the upper floors of buildings. This movie is available at the following URL: http://www.youtube.com/watch?v=nNZxGq70QU)

areas adjacent to the flooded area or to higher ground within the flooded area (Fig. 10.5d). However, it should be noted that this is an interview survey, so there is a certain degree of error in the responses, and data on the dead are not included in the interview survey.

10.3 Preparing for COVID-19 and Future Catastrophes

10.3.1 Utilization of Human Flow in COVID-19

In April 2020, COVID-19, or the declaration of a state of emergency due to coronary infections, was issued in Japan, resulting in a series of lockdowns of cities around the world, causing havoc. In Japan, too, people were encouraged to take actions to avoid

Fig. 10.6 Daily human flow change information posted on the government website. (Cited from https://corona.go.jp/dashboard/.Percentage change at 15:00 on Sunday, February 21, 2021, based on NTT DOCOMO mobile spatial statistics)

three densities, and the situation of human flow in the streets and the number of infected people were discussed in the daily news. Figure 10.6 is a government website that uses data from several mobile communication companies to show daily how much the number of people staying in a representative point in each city has changed from the previous day or before the corona. In this way, the fact that several private companies began to provide human flow data using mobile communication, which started after the Great East Japan Earthquake as described in Chap. 7, at the time of COVID-19 on a business level is a sign that the use of human flow data has penetrated and been accepted by society. Similarly, as mentioned in Chap. 7, it is also significant that we can now see the relationship between people's contact status and the reproduction numbers indicated by the spread of infection.

In addition to the human flow, it is worth mentioning that in Tokyo, an easy-to-understand website on the number of coronary infections was constructed on an open-source basis by Code for Japan, a citizen's group of IT engineers, and it spread to other regions. It is also noteworthy that Code for Japan, a citizen's group of IT engineers, established an open-source-based system that spread to various regions.

10.3.2 Improvement of Measurement Technology for Natural Disasters

Based on the lessons learned from the Great East Japan Earthquake, various additional disaster measurement technologies have emerged. For example, according to the Ministry of Land, Infrastructure, Transport and Tourism's Real Time Naufus website (http://nowphas.mlit.go.jp/info.html), there were 12 GPS wave gauges installed 10–20 km offshore to capture the three-dimensional movement of buoys, observe waves and tide levels, and report any abnormalities 10–20 min before a tsunami strikes. According to the Ministry of Land, Infrastructure, Transport and Tourism's Real Time Nowfus website, there were 12 GPS wave gauges at the time of the earthquake and tsunami, but later, as of March 2014, there were 17, and the number is expected to increase in the future.

In addition, localized flooding due to guerrilla rain has been increasing recently, and high-resolution radar rain gauges are now available. Figure 10.7 shows the rainfall situation using the X-band radar of the Ministry of Land, Infrastructure, Transport and Tourism. Compared to the conventional C-radar, which has a delay of 5–10 min and a 1 km mesh (coverage area of about 120 km), the new radar can provide information at a 250 m mesh level with a delay of 1–2 min.

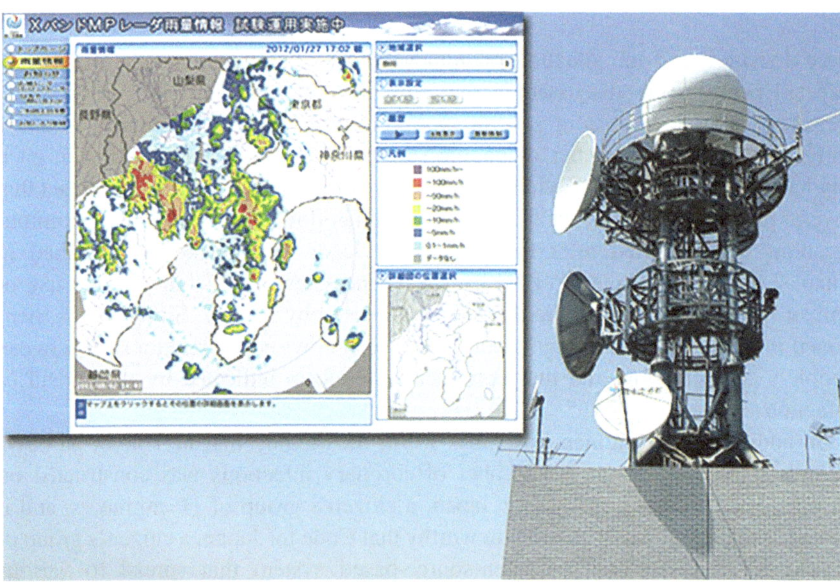

Fig. 10.7 High-resolution rainfall monitoring by X-band radar. (Cited from the Fuji Erosion Control Office, Chubu Regional Development Bureau, Ministry of Land, Infrastructure, Transport and Tourism: http://www.cbr.mlit.go.jp/fujisabo/jimusyo/fujiazami/fujiazami_79/fa7905.html, *Not currently available)

10.3.3 Improvement of Simulation Technology

Even though sensor technology has become more sophisticated, it is not always possible to cover the entire area due to cost and other issues, and given the need for recent forecasts, simulation technology through modeling is still important. In recent years, various types of observation data have become available as historical data, and the models themselves have become more detailed and disaster simulations more sophisticated. For example, in the case of tsunamis, it is possible to model the inundation depth and the outflow of individual buildings, and to take some kind of protective measures in areas where the inundation depth exceeds a certain level when a tsunami is expected in the future. Figure 10.8 was created by Prof. Koshimura and his colleagues, who visually read the damage of individual buildings in Miyagi Prefecture at the time of the Great East Japan Earthquake from aerial photographs provided by the Geospatial Information Authority of Japan, and estimated the damage function based on the inundation depth data of individual buildings (Osaragi and Morisawa 2011).

In addition, especially in the event of an earthquake, buildings may catch fire and spread or collapse, blocking roads and spreading secondary disasters. For this reason, fire spread simulations are often conducted based on the age, material, and density of each building, and blockage simulations are often conducted based on the width of the road. For example, Prof. Osaragi and his colleagues (Osaragi and Morisawa 2011) show a simulation of the Setagaya Ward by borrowing GIS data of individual buildings from the Tokyo Metropolitan Government and visualizing the situation for each district, showing that the northeastern part of the ward is generally more dangerous.

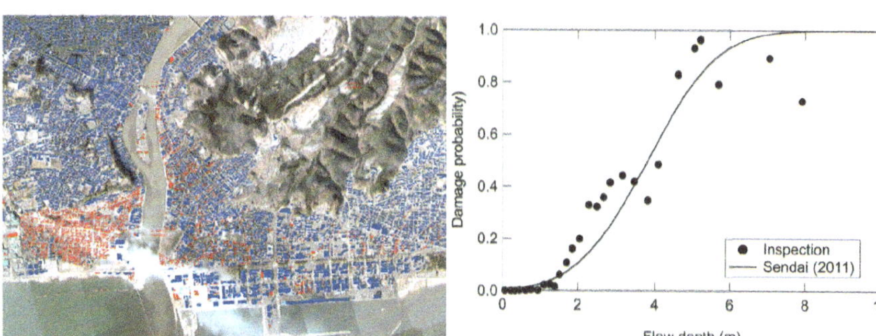

Fig. 10.8 Estimation of damage functions based on tsunami damage to buildings and tsunami inundation depth. (Cited from http://www.bousai.go.jp/kaigirep/chousakai/tohokukyokun/7/pdf/sub8.pdf)

10.4 Conclusion

Japan is a country where various disasters occur, and we have looked at the future of geographic information from the aspect of information technology that can comprehensively respond to these disasters, including the history of the Great Hanshin Earthquake, the Great East Japan Earthquake, and the recent COVID-19.

References

Abe H (2014) Earthquake disaster big data. NHK Publishing

Building Research Institute, Ministry of Construction (1996) Hyogo-ken Nanbu Earthquake damage survey final report, March 1996. Building Research Institute, Ministry of Construction

Cabinet Office (2011) Technical committee on earthquake and tsunami countermeasures based on the lessons learned from the Great East Japan Earthquake, 7th meeting, materials provided by Associate Professor Koshimura. Cabinet Office

Furuhashi D (2011) What we can do—earthquake disaster info sinsai.info—"spatial information infrastructure to support local communities" (supervised by Y. Sekimoto). Nihon Kahyo Shuppan, Tokyo, pp 205–209

Hyogo Prefecture (2008) Determination of damage from the Great Hanshin-Awaji Earthquake, 2008. Hyogo Prefecture. http://web.pref.hyogo.jp/pa20/pa20_000000015.html

Kobe City (2011) Overview and reconstruction of the Great Hanshin-Awaji Earthquake, January 2011. http://www.city.kobe.lg.jp/safety/hanshinawaji/revival/promote/img/honbun.pdf

Ministry of Internal Affairs and Communications (2011) White paper on information and communications 2011. Ministry of Internal Affairs and Communications

Murao O (1999) Study on building damage assessment based on actual damage data of the Hyogo-ken Nanbu Earthquake. Doctoral dissertation, University of Tokyo

Osaragi T, Morisawa T (2011) Simulation model of wide-area evacuation behavior under a large earthquake considering various states and attributes of people staying and moving in a city. Proc Arch Inst Jpn 76(660):389–396

Sekimoto Y, Nishizawa A, Harutoshi Yamada R, Shibasaki J, Kumagai T, Kashiyama T, Sagara Y, Kayama SO (2013) Promotion of data distribution by constructing survey archives for the Great East Japan Earthquake reconstruction support. GIS-The Appl 21(2):1–9

Chapter 11
Utilization in Agriculture, Forestry, and Marine Management

Masahiko Nagai and Yasuhiro Kawahara

Abstract Remote sensing technology is used to monitor large areas in agriculture, forestry, and ocean management. Remote sensing makes it possible to measure the color and temperature of land and sea surfaces, and to know the status of agricultural products and water resources.

Keywords Agriculture · Forestry · Ocean management · Remote sensing · Ship monitoring · Coral reefs

11.1 Agriculture

11.1.1 Growth Monitoring

In agriculture, checking the growth of crops is done through daily patrols by farmers. They observe crop height, leaf color, number of stems, pests, and weeds. Figure 11.1 shows the damage to rice crops caused by the agricultural pest, Brown Planthopper. However, there are concerns that it will become difficult to sustain the cultivation of crops using traditional farming methods, as the number of people working in agriculture declines and the population ages. In addition, there are high expectations for ICT in agriculture (smart agriculture) under the circumstances of increasing abandoned land and declining income from production agriculture.

Satellite remote sensing is a technology that is attracting attention in smart agriculture. Satellite remote sensing can observe the size, shape, number, and nature of objects from space without directly touching them. These features make them

M. Nagai (✉)
Graduate School of Sciences and Technology for Innovation, Yamaguchi University, Ube, Yamaguchi, Japan
e-mail: nagaim@yamaguchi-u.ac.jp

Y. Kawahara
Division of Art and Sciences, The Open University of Japan, Chiba, Japan
e-mail: kawahara2@ouj.ac.jp

Y. Sekimoto, Y. Kawahara (eds.), *Utilization of Geospatial Information in Daily Life*, New Frontiers in Regional Science: Asian Perspectives 65,
https://doi.org/10.1007/978-981-19-6905-8_11

Fig. 11.1 Damage caused by brown planthopper

very suitable for agricultural support. Earth observation satellites are equipped with sensors that measure electromagnetic waves in the visible and near-infrared regions. Plants absorb light at red wavelengths (visible light) and use it for photosynthesis, but cannot absorb wavelengths in the near-infrared region and reflect them back. This is due to the infrared reflection of chlorophyll contained in chlorophyll, which is necessary for photosynthesis and has a strong correlation with the activity level of plants. By utilizing this reflective property, it is possible to diagnose the growth status of crops and determine the appropriate time for harvesting.

In large agricultural areas in Hokkaido, large agricultural machinery is used to harvest rice and wheat, and large quantities of crops are transported at a time to processing facilities. Satellite remote sensing technology is helping to improve the efficiency of agricultural operations, because if the appropriate harvest time can be determined from the growth of the rice and wheat, it is possible to plan the order of harvesting by agricultural machinery and the efficient use of facilities.

Aomori Prefecture uses remote sensing to estimate the protein content of rice. Protein is one of the factors that determine the taste of rice, and rice is said to be tasty when it is low in protein. By correlating rice activity data observed from satellites with actual measurements of protein, the taste of rice can be predicted before harvest, and rice with low protein content can be carefully selected and harvested to promote rice branding.

11.1.2 Orchards and Golf Courses

Farmers operating large orchards must manage thousands of fruit trees, and managing them from the ground requires an enormous amount of labor and time. Remote sensing technology makes it possible not only to detect the number of trees, but also to estimate canopy size, optimize efficient tree planting, and detect diseased trees based on spectral information from satellite data. Satellite remote sensing can measure plant activity using near-infrared wavelengths. By quantifying plant activity, called the vegetation index (NDVI), it is possible to detect the location of fruit trees that have been damaged by pests and diseases.

Golf courses are vast, and it is difficult to detect unhealthy lawns only by monitoring on the ground. For this reason, golf courses have traditionally required a great deal of money, labor, and time for turf management. However, remote sensing technology, which combines regular observations by satellite images and detailed surveys by drones, is expected to significantly reduce costs by quantifying the condition of the lawn, preventing turf from dying due to lack of water, and effectively applying appropriate fertilizers and chemicals. This technology is expected to significantly reduce costs (Fig. 11.2).

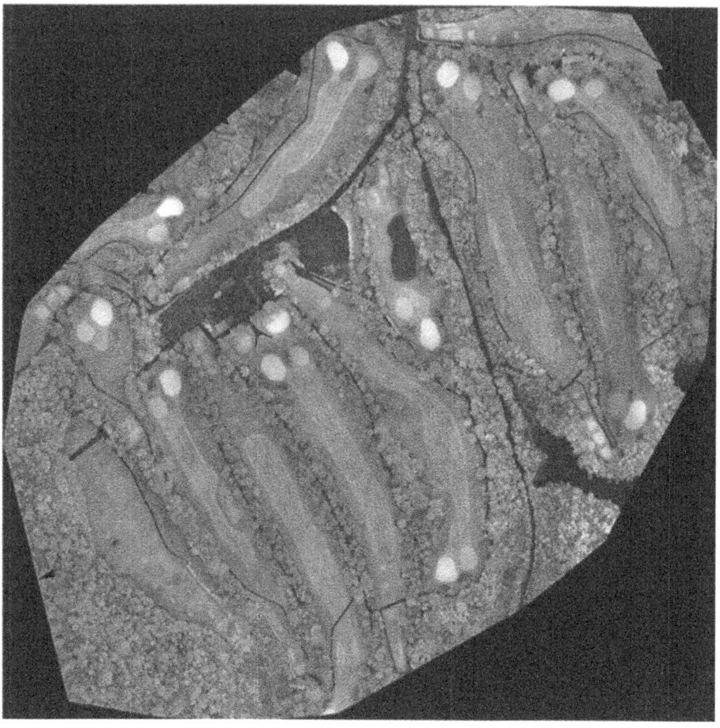

Fig. 11.2 Golf courses observed by drones

11.2 Forestry

11.2.1 Forest Monitoring

Forest information in local governments (prefectures and municipalities) is managed by GIS (Geographic Information System). Each forest owner has a forest division boundary in each polygon and has attribute information such as tree lumber volume estimated by tree age and location conditions. However, many forest owners do not know the current status of their land and GIS data is not updated. Therefore, in many timber production sites, field surveys are used to estimate actual timber volume.

Remote sensing technology has been deployed to monitor and assess the conditions of the environment and resources in large areas of forested lands. In vast forests, deforestation, seasonal changes, and periodic information updates for forest management are difficult to achieve with ground-based surveys alone. An effective method of forest survey is aircraft laser measurement. Laser surveying can measure tree height and canopy shape by analyzing point cloud intensity. Terrestrial laser scanning (TLS) and photogrammetry by drone and laser surveying are becoming increasingly popular for forest surveys in small forest areas.

The method used to monitor forests by satellite remote sensing is land cover classification using the reflectance properties of visible and near-infrared light. Based on satellite data, land can be classified as urban, agricultural, water, bare, or forest. Furthermore, this forest classification can be further refined to include hardwoods, conifers, mixed forests, and bamboo forests (Fig. 11.3).

11.2.2 Bamboo Forest

The area of bamboo forests is increasing nationwide. As demand for bamboo materials declines, bamboo groves in many areas are no longer managed, and the

Fig. 11.3 Drone image for forest (left) detects only conifers (right)

number of vigorous and fast-growing bamboos is increasing as they grow wild. There is also concern about the impact on the surrounding environment and ecosystems caused by the expansion of naturally neglected bamboo forests, and efficient management of the increased bamboo forests is urgently needed. However, the current situation is that the creation of vegetation maps for managing bamboo forests is a very labor-intensive and time-consuming process that is carried out through field surveys.

Satellite remote sensing is expected to be an effective means of monitoring fast-growing bamboo forests because it can observe large areas at once from space and the same areas can be repeatedly observed over and over again. Once satellite data becomes available, bamboo groves can be easily monitored in a short time. To detect bamboo groves, areas covered by bamboo groves should be identified in advance by field surveys, and satellite data should be used as teacher data for classification. For more detailed classification, it is also important to take into account various factors such as differences by species, mixed forests of bamboo and other trees, seasonal changes, and regional characteristics (Fig. 11.4).

Fig. 11.4 Abandoned bamboo grove

11.3 Maritimes Management

11.3.1 Vessel Monitoring

- Satellite remote sensing is used to monitor vessels in the vast ocean. Examples of satellite data applications in the ocean include monitoring sea surface temperature, currents, ocean winds, ice floes to aid in vessel navigation, oil spills, and vessel monitoring.
  Unlike observations on the ground, observations at sea can easily detect vessels because there are fewer complex structures and landmarks. Figure 11.5 shows the detection results of ships in Tokyo Bay observed by ALOS-2 with a resolution of 3 m. The ships are seen in white on the sea surface (black area).
  Vessel monitoring uses data acquired from satellites as well as data acquired by a device called AIS (Automatic Identification System), which continuously sends information on vessel identification, such as name and type of vessel, as well as position, speed, and other information related to the vessel's course. Vessels that meet certain standards are required to be equipped with AIS. By comparing AIS information with vessels detected by satellite, it is possible to identify vessels that do not use AIS and thus detect suspicious vessels.
- Although monitoring suspicious vessels from space is technically at the stage of practical application, it is difficult to use it on a regular basis at present because the frequency of observation is not sufficient. In order to increase the frequency of observation, a technology called constellation, in which multiple satellites are launched and placed in the same orbit, is being considered. If the frequency of observation is increased in the future, satellites can be used effectively for safe navigation and security.

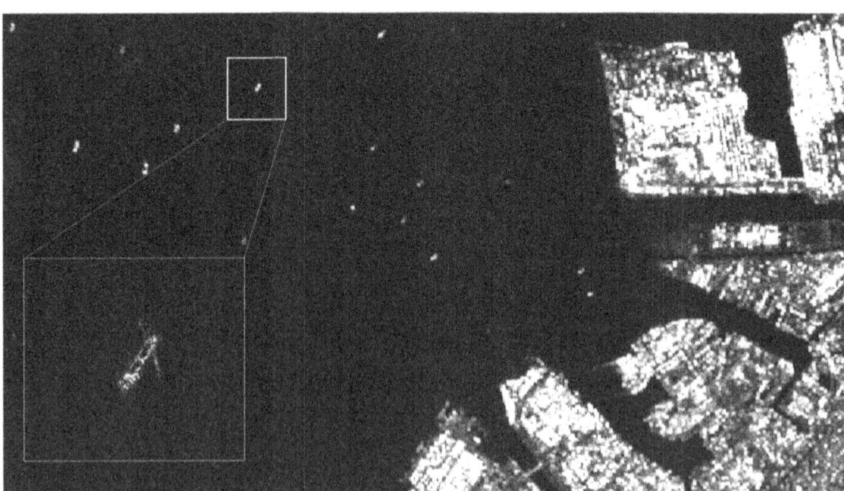

Fig. 11.5 Detection of vessels by the Daichi-2 (© Yamaguchi University)

11.3.2 Coral Reefs

- Coral reefs play the most important role in marine ecosystems due to their high biodiversity and productivity. In recent years, however, the destruction of coral reefs has become a major problem worldwide. Some of the causes include eutrophication of coastal areas due to agricultural soil runoff, damage caused by excessive tourism use, and the bleaching of coral reefs due to rising sea water temperatures caused by global warming.
- Appropriate protection of marine environments such as coral reefs requires observations from an ecological perspective, using diving surveys and underwater cameras in coral reef areas. However, these localized monitoring requires a huge amount of time and cost to repeatedly monitor a wide range of seafloor environments over a long period of time.
- Satellite remote sensing can be used to repeatedly observe a wide area of the ocean as well as land areas. To observe coral reefs in the ocean through remote sensing, it is necessary to understand the wavelength of light in the ocean. The condition of coral reefs is indirectly inferred by changes in the color of the sea surface over the coral reefs.

Technological developments in sensors mounted on satellites have made it possible to acquire detailed information with high resolution. Observation using multiple wavelength bands has become possible, and sensors with shorter wavelength bands are being used to monitor underwater coral reefs. Figure 11.6 is a satellite image of the area around Taketomi Island, Okinawa Prefecture. In the vicinity of Taketomi Island, coral reef bleaching due to environmental changes such as rising seawater temperatures has become a major problem. Although there are limitations of environmental factors such as coral reef seawater transparency, waves, and deep water, it will be possible to observe coral reefs over a wide area of ocean.

11.3.3 Fish Group Exploration

In fisheries, fish finders using ultrasonic sensors are used. As the name suggests, a fish finder is a machine that searches for schools of fish, and by installing sensors on the bottom of the vessel, it can also measure the depth and undulations of the seafloor. The configuration of the fish-finding equipment is shown in Fig. 11.7. The system consists of a display/control unit, an ultrasonic transmitter/receiver unit, and a battery. The transmitter/receiver unit is installed on the bottom of the ship or other areas that can be probed to measure reflected ultrasonic waves on the seafloor or fish. The ultrasonic transmitter/receiver emits pulses of ultrasonic waves to the seafloor and measures the time taken to receive the reflected waves. The speed of ultrasonic waves traveling through water is 1500 m/s. When, for example, it takes 2 s to measure the reflected wave from a transmitted ultrasonic wave at the seafloor, the distance (depth) to the seafloor is 1500 m. Ultrasonic frequencies in the bandwidth

Fig. 11.6 Coral reefs around Taketomi Island, Okinawa (© Yamaguchi University)

from 15 kHz to 200 kHz are often used. Using the fact that low frequency waves have a large directivity angle (the angle at which the traveling waves spread out), and higher frequency waves have the opposite, for example, 50 kHz ultrasonic waves (directional angle: about 50°) are used to search for fish schools over a wide area of ocean, and then 200 kHz ultrasonic waves (directional angle: about 12°) are used to search a smaller area with high accuracy. Figure 11.7 shows an example of the display screen of a fish finder. There are two screens, one on the left and the other on the right, showing the measurement results of ultrasonic waves of different frequencies. In each screen, the horizontal axis is the time axis, and the rightward direction represents newer information. The vertical axis represents the depth, and the measurement result is indicated by the coloring of the area at the depth where the ultrasonic waves are reflected by the object. Colors above it, such as dots, indicate

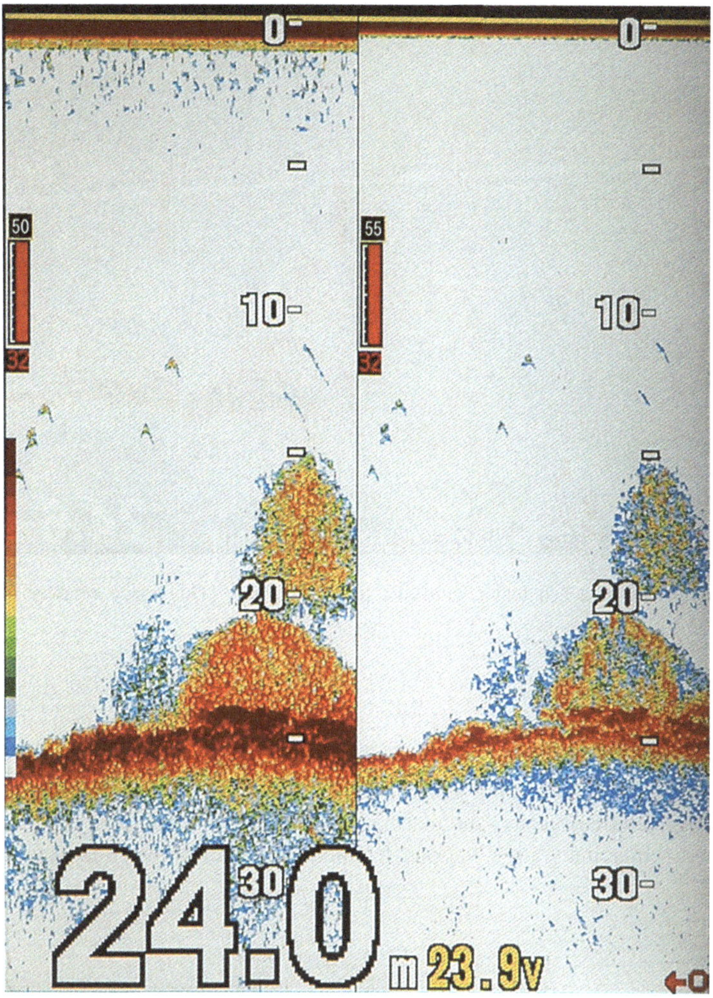

Fig. 11.7 Display screen of a fish finder. [This image courtesy of Honda Electronics Co., Ltd.]

the presence of fish. The shape and size of the fish schools that appear in the measurement results are used to predict the species of fish, which is useful for fisheries.

On fishing vessels, GPS plotters are used with fish finders. This is a machine that records the position measured by GPS and draws a navigational track on a map. By using this machine with a fish finder, it is possible to visually recognize when and where the fish finder responded, allowing efficient fishing operation. Figure 11.8 shows an example of the screen of a fish finder with GPS plotter function. The center of the map on the left screen indicates the location where the measurement is conducted.

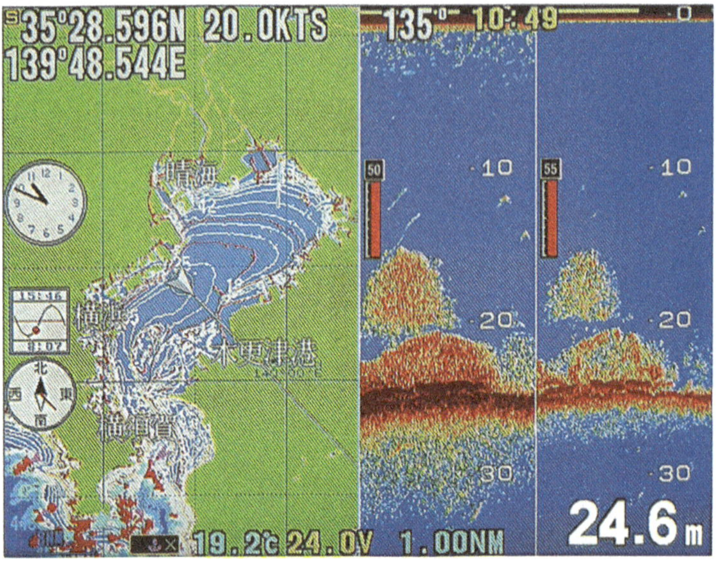

Fig. 11.8 Screen of a fish finder with GPS plotter function. [This image courtesy of Honda Electronics Co., Ltd.]

11.4 Summary

In this chapter remote sensing technology has been described for monitoring large areas in agriculture, forestry, and marine management. We have also explained how it is used while integrating various technologies such as satellite data, drones, and GIS.

Chapter 12
Utilization in Behavior, Ecology, and Cultural Property Surveys

Yasuhiro Kawahara

Abstract The miniaturization and power savings of information and communication devices and sensors have made it possible to record behavioral and physiological information together with geospatial information using devices that people carry all the time and sensors attached to animals. This information has been used for behavioral recognition and ecological research. In addition, to estimate the lifestyle and ecology of past eras, geospatial information was created from archaeological sites and archived in the field of archaeology. This chapter describes the technology and forms of information used to realize these objectives, along with case examples.

Keywords Mobile sensing · Information communication terminal · Biologging · Biometric sensor · Archaeological map · Underground radar survey

12.1 Mobile Sensing

12.1.1 Mobile Computing

The miniaturization and low-powered functioning of information and telecommunications equipment have made it possible for many people to carry information communication terminals, such as smartphones, at all times. The fact that people carry small computers connected to such a network at all times makes it possible to keep track of the information of those carrying them in real time. Figure 12.1 shows the conceptual position of information and communication terminals in the living environment and their components. Information about the holder, the surrounding environment, and the network are linked through the external interface in the composition of the information communication terminal in the figure. External interfaces for cellular phones, smartphones, and other mobile devices include

Y. Kawahara (✉)
Division of Art and Sciences, The Open University of Japan, Chiba, Japan
e-mail: kawahara2@ouj.ac.jp

Y. Sekimoto, Y. Kawahara (eds.), *Utilization of Geospatial Information in Daily Life*, New Frontiers in Regional Science: Asian Perspectives 65,
https://doi.org/10.1007/978-981-19-6905-8_12

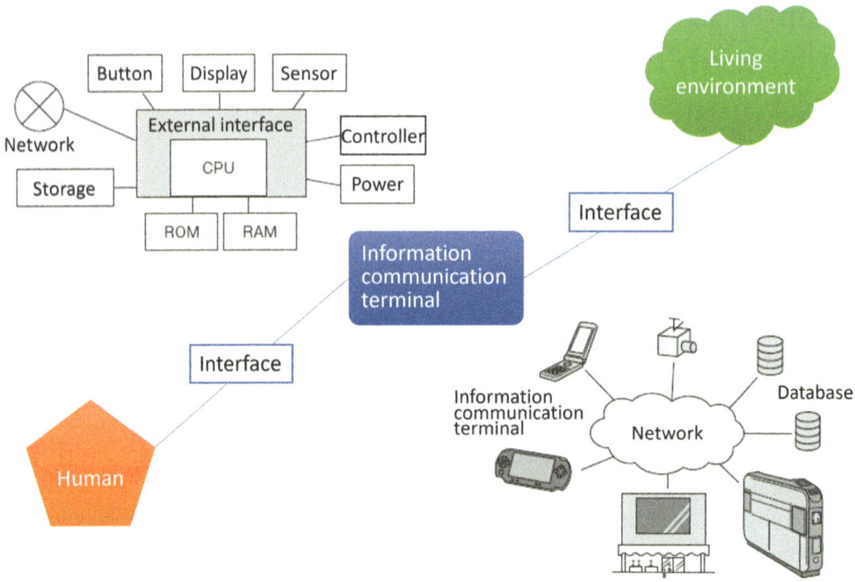

Fig. 12.1 Information communication terminals in the living environment

operating units such as buttons and touch panels, display units such as displays and LEDs, sensor modules such as acceleration and illumination sensors, cameras, microphones, and other devices. Information on human behavior and the surrounding environment obtained through these external interfaces is processed by the mobile information communication terminal's computing unit (CPU), stored in the terminal itself or on a server on the network, and used in various services.

In Japan, many cell phone terminals have been equipped with GPS functions since 2009, and many of the currently popular smartphones are equipped with wireless LAN and near-field communication (NFC) functions. Using these functions, the location of the information communication terminal can be determined by combining the information obtained from the external interface with the location information of the terminal. It is possible to estimate when and where a particular behavior or environmental transition has occurred. Under these circumstances, there are various services to users of information communication terminals that use location and external interface input information. Some organizations constantly accumulate user behavior and provide information or services that are suitable for that information. Recording one's activities and circumstances in daily life and organizing them in a database is also known as lifelog.

Today, wearable terminals, such as smartphones and smartwatches, are becoming popular. The use of wearable information communication terminals to collect and process information is known as wearable computing. In addition, information communication devices in a living environment (e.g., motion sensors, digital signage, wireless LAN access points, information appliances, and street cameras) can reinforce and communicate user information obtained from mobile terminals. Such processing by information communication terminals installed in a way that fits daily life is called ubiquitous computing. In combination with wearable computing, it enables more precise mobile sensing. Wearable computing presents several challenges. The low power consumption of terminals, the application of efficient and fast information communication methods, and the development of embedding technology for flexible sensor function materials will enable better system designs that do not impose burdens on daily users.

12.1.2 Human Sensing

Let us consider the sensing of human information (human sensing) that the aforementioned mobile computing can realize. Mobile information communication terminals have multiple ways of determining their locations. The most commonly used functions are GPS, wireless LAN, and NFC. GPS can determine the location of a terminal with an accuracy of several meters where four or more GPS satellites are visible from a terminal equipped with GPS functions. Positioning is not possible indoors or in areas with a high concentration of high-rise buildings. GPS-based positioning services are standardized platforms on a global scale that can be used with the same interface anywhere in the world.

Using the wireless LAN function of a mobile information communication terminal, the terminal's location can be estimated from the signal strength and ID of the surrounding wireless LAN access points. This positioning service requires the service provider to know the IDs and locations of access points or to have learned the IDs of access points that can be received at the positioning location. If these conditions are set, terminal location estimation can be as accurate as or more accurate than the interval between access point installations.

The location estimation method for mobile terminals using the NFC function is simple; the location of the NFC reader/writer terminal that communicates with a mobile terminal or card with an embedded NFC chip becomes the terminal location at the time of communication. Examples of NFC reader/writer terminals include digital signage and automatic ticket gates for transportation. This method is equivalent to the method used to monitor the distribution and inventory of RF-tagged products. The presence of NFC reader/writer terminals at locations where they are used daily, such as automatic ticket gates at train stations, also makes it possible to monitor the flow lines of the users.

A sensor module embedded in an information and communication terminal can collect information of user activity. Accelerometers are frequently used for this purpose. Figure 12.2 shows an example of the acceleration and atmospheric pressure measured with a mobile sensor while moving around a commercial building. There is a movement to use such data to estimate behavior in the city, ascertain lines of flow in specific areas, and provide useful information to individual users. The use of wearable information communication terminals on specific parts of the body enables

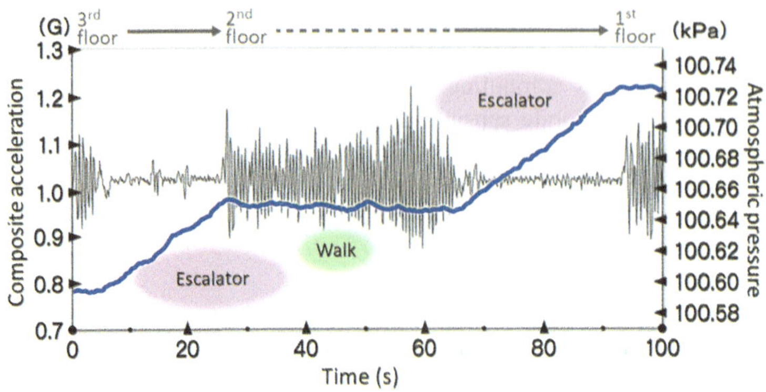

Fig. 12.2 Measurement of acceleration and atmospheric pressure by mobile sensors

the sensing of a wider variety of biometric information. For example, the changes in heart rate and autonomic nervous system activity indices can be calculated from electrocardiograms continuously sensed by a 15 g sensor affixed to the chest. Thus, it is also possible to collect useful mental and physical healthcare information.

In addition, other biometric sensors that allow measurement while being carried out include eye-tracking meters, respiration sensors, electromyographs, electroencephalographs, and near-infrared spectroscopy (NIRS) devices. Although there are still issues to be solved for daily use, such as eliminating noise caused by body movements and further reducing size and power consumption for daily wear, research is also progressing to decipher user intentions and assumptions by analyzing brain functions using such devices. These sensors may be incorporated into mobile terminals for daily use in the future. In particular, the interface between the brain and machines (information devices) is called a brain-machine interface (BMI). The design of a BMI that fits in daily life will lead to the development of services that estimate and utilize user will and attention.

12.1.3 *Human Probes*

A mobile information communication terminal can be used to sense the user's surrounding environment. By sensing the light, sound, and thermal environments, it is possible to monitor the user's surrounding environment. Using the system in combination with location information devices such as GPS to grasp the user's surroundings and the user's actions caused by the surroundings, it is possible to obtain a wide range of environmental information. This method of monitoring environmental information through scanning by human movement is often referred to as human probing. It is useful for simple information surveys that do not change over time (e.g., radiation levels, topography, etc.). By aggregating information provided by a large number of users at different times and locations, the environmental information for a wide area can be visualized. A service that aggregates and releases weather status reports via information communication terminals has realized detailed live-weather conditions.

12.2 Biologging

12.2.1 Recording Wildlife Information

Global positioning system (GPS) and mobile sensing devices have also been used to survey animal ecology. A GPS receiver was attached to the animal to track animal movements, and movement track data were collected through receiver retrieval or wireless communication. In the case of power saving by data loggers attached to animals or when GPS cannot be used to monitor animals living underwater where GPS satellite signals do not reach, there is a method to estimate the time of sunrise/sunset and the length of the day by using the output of the illuminance sensor and the time to determine the position of the animal (longitude can be determined from the time of sunrise/sunset and latitude from the length of day). It is also possible to estimate the locus of movement by attaching a radio wave transmitter to the animal and measuring the direction and intensity of the transmitted radio wave using a directional antenna and radio wave receiver.

An animal's movement status and posture can be estimated by attaching and measuring an acceleration sensor to the animal. Whether an animal is running or stopped can be estimated by extracting the characteristic acceleration waveform of each state, and posture (standing, running, sleeping, etc.) can be estimated by detecting the gravitational acceleration measured by the acceleration sensor. For example, an animal that walks on all four limbs should show accelerometer values close to the gravitational acceleration of 9.8 m/s^2 in the same direction as the limb bar while walking or standing. Animal behavior information, such as movement status and posture, is managed together with spatiotemporal information expressed in time and location and analyzed to determine where and when a particular individual is in a particular condition. Biometric information requires the implantation of sensors in the body or the use of electrodes to measure bioelectrical signals. By attaching a camera, microphone, thermometer, hygrometer, and illuminance meter to the animal, information about the animal's surrounding environment can be recorded. This is the same type of measurement as that of the human probe mentioned in the previous section (Table 12.1).

The animal's behavior, physiological condition, and surrounding environment, measured and estimated by these methods, determine the animal's specific behavior. For example, when and where target recognition, foraging behavior, spawning timing, communication with mates, etc., can be determined by combining sensed information.

Table 12.1 Sensors to detect the animal state

	Measurement target	Sensor/measurement system
Movement	Location	GPS, telemetry, illuminator
	Altitude	GPS, telemetry
	Depth in water	Strain gauge
Behavior	Move/stop	Accelerometers, gyroscope
	Posture	Accelerometers
	Direction	Gyroscope, geomagnetic sensor
	Body temperature	Thermometer
Physiological information	Heart rate	Cardiometer
	Myoelectricity	Electromyograph
Surrounding environment	Temperature and humidity	Thermohygrometer
	Illumination	Illuminance meter
	Image	Camera
	Sound	Microphone

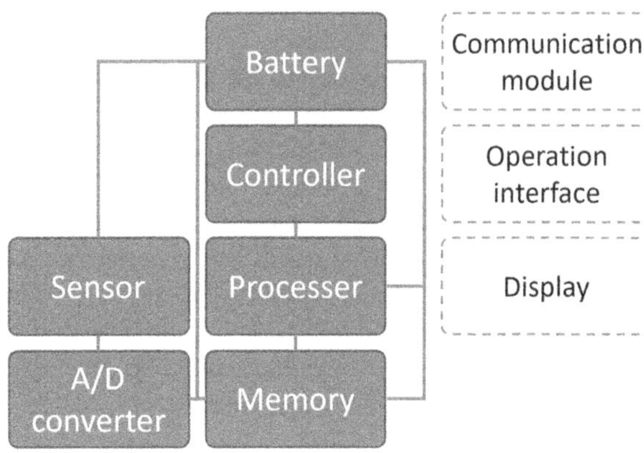

Fig. 12.3 Data logger configuration

12.2.2 Data logger

To measure animal information, it is necessary to attach sensors to the animals and record their output values. This device is called a data logger. As shown in Fig. 12.3, the basic configuration of a data logger consists of a sensor, analog-to-digital conversion (A/D) converter, processor, controller, memory, and battery with additional communication modules, operation interfaces, mounting tools, and display devices added as necessary.

Sensors are used to capture real-space events as digital data on information devices. Sensors are devices that convert physical phenomena and chemical substances into electrical signals, and the converted electrical signals (analog signals) are converted into digital signals by analog-to-digital (A/D) converters (Fig. 12.4). The conversion of real-space events into electrical signals using sensors is called sensing, and the acquisition of events as data using these devices is called sampling.

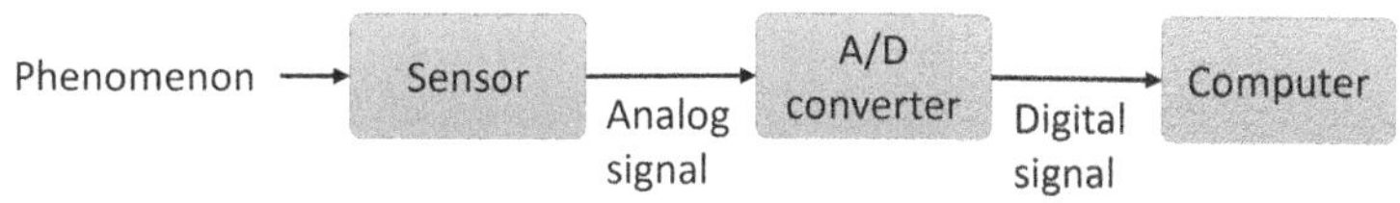

Fig. 12.4 Digitization of events in real space

The sampled data are stored in the data logger's memory and retrieved externally through the communication interface as needed.

The most important factor to consider when attaching a data logger to an animal is its weight. The weight of a data logger on an animal living on land or in the air is 2–5% of the animal's body weight. Therefore, the data loggers must be designed under the condition that the weight, including the mounting jig, must be less than this weight. When sampling over long periods and at high sampling frequencies, the power consumption during the entire biologging period increases, and a battery of appropriate capacity must be built in, increasing the weight of the system. Because wireless communication requires a relatively large amount of power, to reduce the weight of the battery, communication timing is considered to keep the communication time to the minimum necessary, and a communication protocol that enables power saving is selected. Data from data loggers attached to wild animals must be retrieved for ecological analysis; however, it is difficult to retrieve data or data from data loggers attached to wild animals that are not known when and where they are, and various efforts have been made to retrieve the data. For example, there are methods such as building a radio wave transmitter (beacon) into the data logger and retrieving it by searching for beacon radio waves with a directional receiver or building a public wireless communication module into the data logger and remotely downloading data when the animal is in a public wireless communication network. Each method requires the removal and retrieval of the data logger from the animal, often by capturing it and considering the animal's habits. One method of removing data loggers from animals without capturing them is to detach the data logger attachment jig by timer control. Because the detached data loggers must be detached at a location where they can be retrieved, the timing of the control of the detachment device should be considered depending on the survey environment and animal species.

12.2.3 Biologging on Land, in the Sea, and Air

The weight of a data logger that can be attached to an animal living on land or in the air is estimated to be 2–5% of the animal's body weight. For example, an elephant weighing 4000 kg can be fitted with a data logger weighing up to 200 kg, but a dog weighing 4 kg can be fitted with a data logger weighing only 200 g. For animals that live in water, the allowable range of data logger weight should be considered, taking buoyancy and swimming resistance into account.

Biologging is a necessary means of elucidating the life (daily life) of wild animals that cannot be directly observed. In particular, the biology of animals living in the

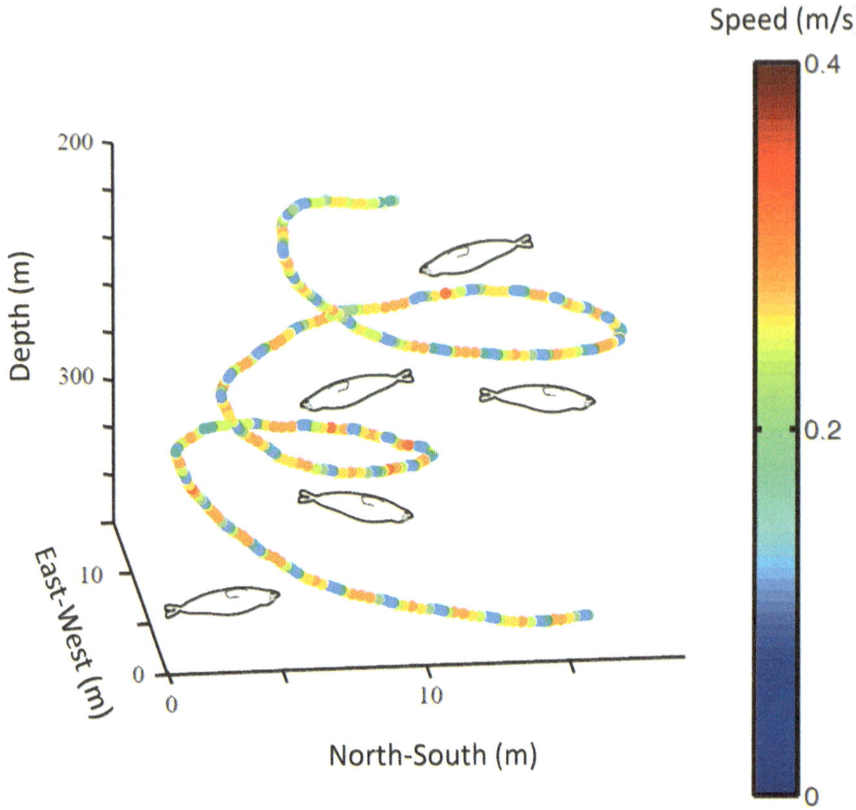

Fig. 12.5 Travel tracks of a northern elephant seal in the water. [This image courtesy of Yoko Mitani (Wildlife Research Center of Kyoto University)]

sky and sea has elucidated animal life that cannot be obtained in laboratories and by outdoor manual observation. For example, satellite-based systems have been used in the ecological surveys of migratory birds (Shimazaki et al. 2004). The satellite receives signals from a transmitter attached to the bird and analyzes them to estimate the location of the bird. Studies have been conducted to identify migratory routes, breeding sites, stopover sites, and wintering grounds of migratory storks, which are listed as endangered species, and help preserve their habitat.

Accelerometers and pressure gauges can measure the three-dimensional position and swimming speed of animals living in the water. Figure 12.5 shows an example of the movement path of a northern elephant seal at a depth of 400 m. It can be assumed that the seal is in a slow spiral dive (resting) in the deep sea, with its body in a belly-up position (Mitani et al. 2010). In addition, recording physiological information and location information will enable a more detailed elucidation of animal behavior. Figure 12.6 shows the daily behavior data of sunfish, including underwater depth, feeding behavior (captured by the camera), body temperature, and water temperature. From these data, the behavior of sunfish feeding at depths of

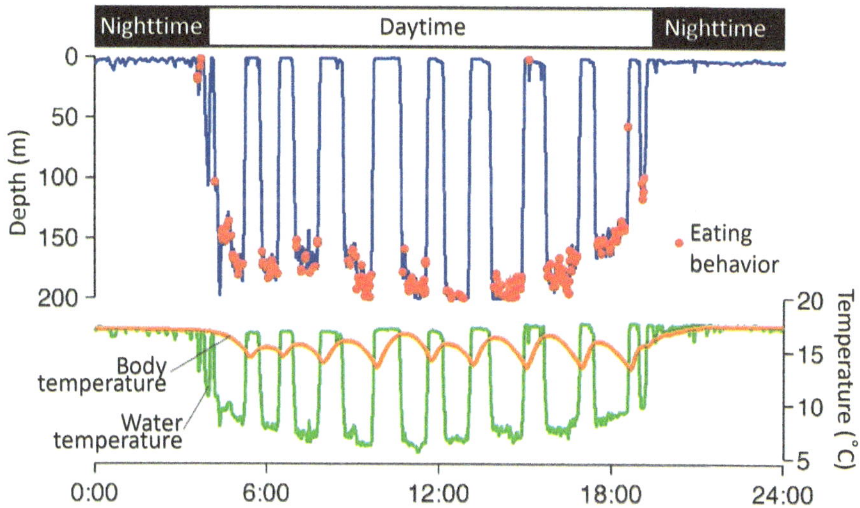

Fig. 12.6 Depth and water and body temperature changes of sunfish over a day. [This image courtesy of Itsumi Nakamura (Organization for Marine Science and Technology, Nagasaki University)]

100–200 m and recovering their body temperature at the water surface when their body temperature cooled was observed (Nakamura et al. 2015).

Some studies on insect behavior sensing have been conducted using telemetry techniques. Telemetry technology, attaching radio transmitters to honeybees, is used to track foraging flights (Riley et al. 2005). The meaning of the "figure-eight dance" used to communicate the location of the feeding grounds to the mate had not yet been hypothesized, but this method of biologging will help prove the hypothesis.

12.3 Use of Geospatial Information in Cultural Property Surveys

12.3.1 Measured Drawings of Archaeological Remains

In archaeology and site surveys, a measured map of the layout of archaeological sites is called a measured drawing. This map is left behind for archival purposes, as shown in Fig. 12.7. This figure depicts columns and other remains of different ground qualities and colors. By combining a plan view and a cross-sectional view, the remains can be understood in three dimensions: In this figure, the coordinates and elevations are filled to provide an absolute reference of the site's location. In the measured drawings of the remains, there are targets, and the layout of the remains is drawn using these targets as indicators. There are also measured drawings that do not set permanent objects in the target, in which case, it is impossible to determine where

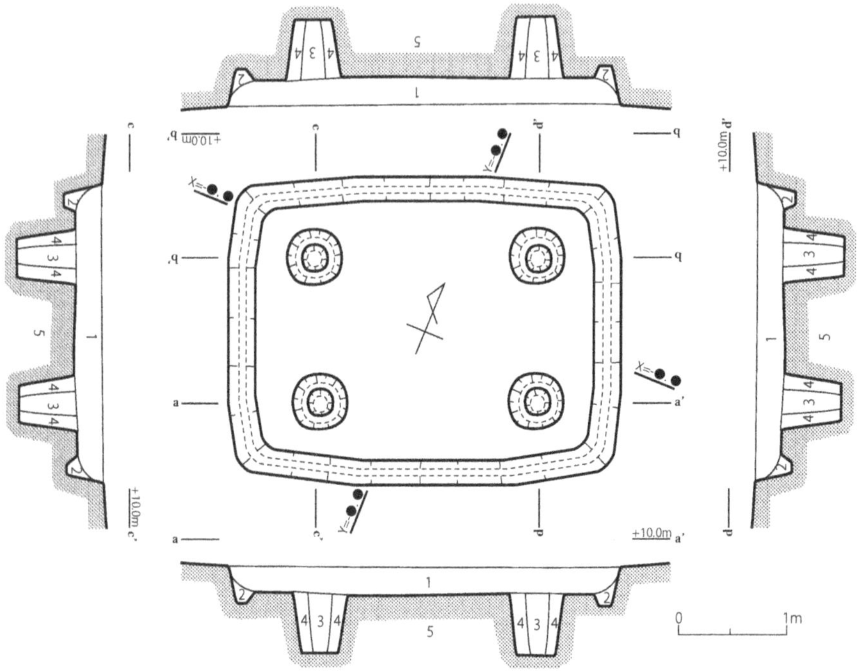

Fig. 12.7 A sample of measured drawings of archaeological remains. [This image courtesy of Keisuke Endo (Amagasaki City Museum of History)]

the remains exist. Today, archaeological sites with geospatial information are archived by defining targets with latitude and longitude coordinates or by managing measured drawings in GIS.

12.3.2 Various Surveying and GIS

It has become common to manage the remains map together with information on current Earth coordinates (latitude, longitude, and elevation) to identify the location of the remains. Most surveys are conducted using conventional methods, such as total stations with distance measurement functions and photogrammetric surveying when mapping archaeological sites. Recent technological advances have led to the introduction of various surveying methods.

One of these is a laser scanner. A laser scanner can acquire information on the 3D topography of the surrounding environment and layout of objects by emitting a laser beam from the device to the surrounding area and performing plane ranging. The surveying method that uses this technique is called LiDAR. Examples of laser scanner photographs and output drawing images are shown in Figs. 12.8 and 12.9.

Investigations of subsurface conditions without excavation work have also been conducted. A radar probe is used to measure the subsurfaces. 3D information,

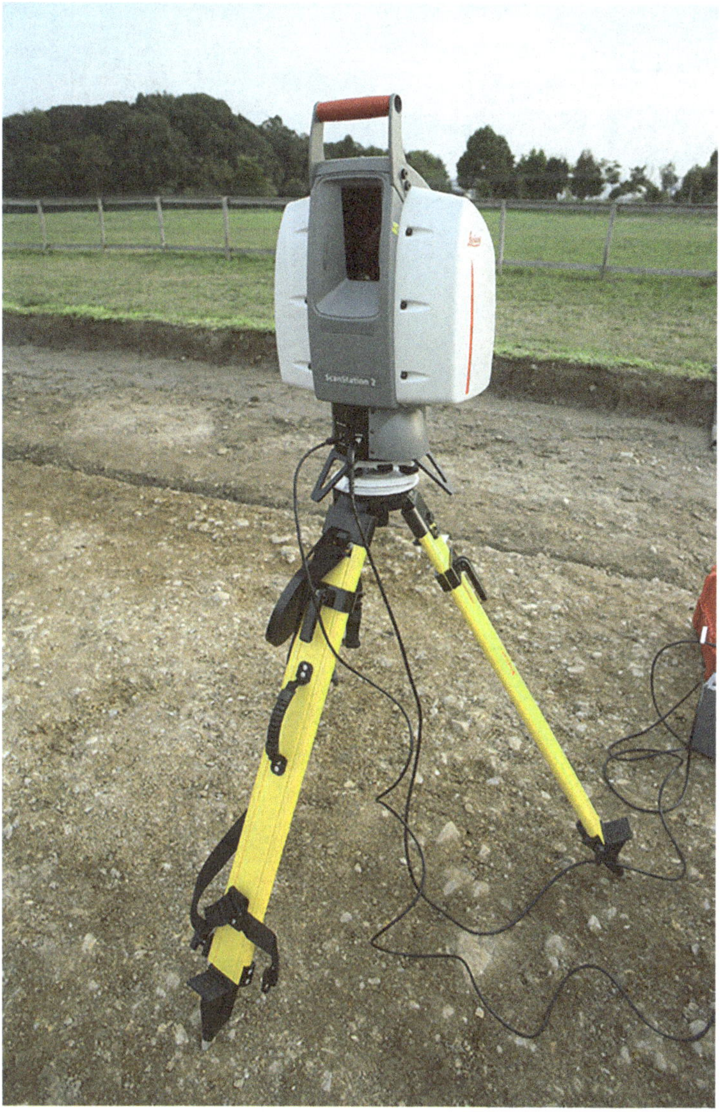

Fig. 12.8 Laser scanner. [This images courtesy of Akihiro Kaneda (Nara National Research Institute for Cultural Properties)]

including depth, can be obtained. In addition, using software that combines images and geospatial information, composite photos with embedded 3D location information can be created from digital camera photos.

This information is being increasingly managed using GIS. Because GIS can overlay different types of maps and images with latitude and longitude information as layers. It is a useful tool for archaeological studies; for example, by overlaying

Fig. 12.9 Depiction of archaeological sites by a laser scanner. [This images courtesy of Akihiro Kaneda (Nara National Research Institute for Cultural Properties)]

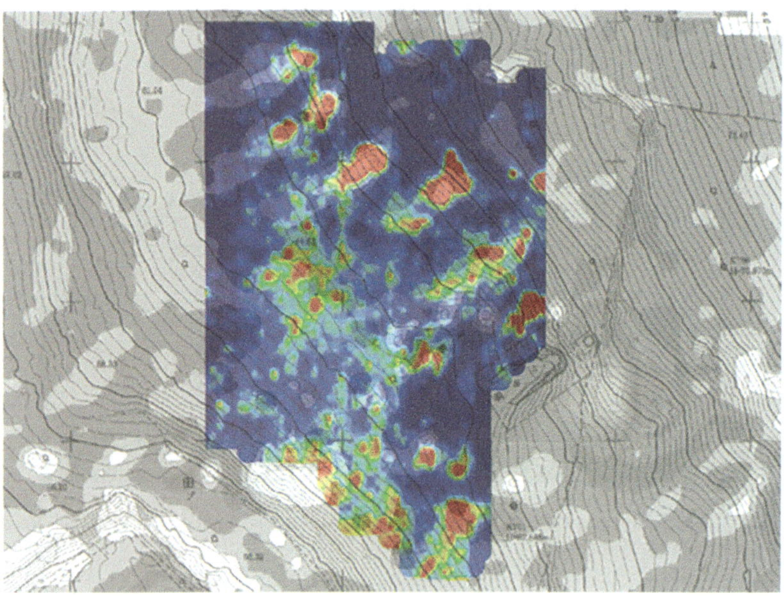

Fig. 12.10 Synthesis of ground-penetrating radar images and topographic maps by GIS. [This image courtesy of Akihiro Kaneda (Nara National Research Institute for Cultural Properties)]

current topographic information with a map of the remains. Figure 12.10 shows a map superimposing a topographic map and ground-penetrating radar image, which can be used to estimate the location of the kiln on the map.

References

Mitani Y, Andrews RD, Sato K, Kato A, Naito Y, Costa DP (2010) 3D resting behaviour of northern elephant seals: drifting like a falling leaf. Biol Lett 6:163–166
Nakamura I, Goto Y, Sato K (2015) Ocean sunfish rewarm at the surface after deep excursions to forage for siphonophores. J Anim Ecol 84(3):590–603
Riley JR, Greggers U, Smith AD, Reynolds DR, Menzel R (2005) Waggle dance controversy resolved by radar records of bee flight paths. Nature 435(7039):205–207
Shimazaki H et al (2004) Network analysis of potential migration routes for Oriental White Storks. Ecol Res 19:683–698

Chapter 13
International Applications

Masahiko Nagai

Abstract In this chapter, examples of the use of geospatial information and remote sensing are introduced such as international cooperation for SDGs and huge disaster in global perspective.

Keywords SDGs · International cooperation · Malaria · COVID-19 · Remote sensing · Disaster

13.1 Global Issues

13.1.1 SDGs and Space Technologies

The Sustainable Development Goals (SDGs) are international goals adopted at the 2015 UN Summit (UN 2021) that set 17 targets to be achieved by 2030 to achieve a sustainable world (Fig. 13.1). The goals are set for poverty, hunger, health, education, gender, water and sanitation, energy, growth and employment, innovation, inequality, cities, production and consumption, climate change, marine resources, terrestrial resources, peace, and means of implementation. In order to achieve these goals, space and geospatial information is expected to be utilized. It is also difficult to evaluate the situation using the same indicators.

Satellite data can be objectively evaluated by the same sensors around the world using the same criteria. If the entire world can be evaluated based on the same criteria, it will become clear where on the globe to do what. In developing countries in particular, satellite data can be used to clarify issues that have been difficult to solve due to a lack of data on the ground, and there are many cases where this has led to solutions to issues that had not been apparent before.

M. Nagai (✉)
Graduate School of Sciences and Technology for Innovation, Yamaguchi University, Ube, Yamaguchi, Japan
e-mail: nagaim@yamaguchi-u.ac.jp

Y. Sekimoto, Y. Kawahara (eds.), *Utilization of Geospatial Information in Daily Life*, New Frontiers in Regional Science: Asian Perspectives 65,
https://doi.org/10.1007/978-981-19-6905-8_13

Fig. 13.1 The 17 sustainable development goals in the SGDs. (Source: United Nations)

As a representative effort to use satellite data for the SDGs, the Japan International Cooperation Agency (JICA) and the Japan Aerospace Exploration Agency (JAXA) have established the "JICA-JAXA Tropical Forest Early Warning System" by monitoring deforestation and changes in tropical forests in about 80 countries using the Earth observation satellite "ALOS-2" and the system monitors illegal logging and changes in forests from space and aims to manage the rich tropics (SDG Goals 13 and 15).

In developing countries, where damage from water-related disasters such as floods and droughts is increasing, hydrological data such as rainfall and river flow rates are in short supply. The data is made publicly available. JAXA is working to reduce flood damage by observing rainfall from space (SDG Goals 6, 11, and 13).

Thus, space data utilization is one of the key solutions for achieving the SDGs and for solving problems.

13.1.2 Global Warming

Global warming is a phenomenon in which the average global temperature is rising due to the increase of greenhouse gases such as carbon dioxide and methane in the atmosphere caused by human activities. As temperatures rise, there are concerns about sea level rise due to sea water expansion and melting of glaciers, as well as the occurrence of natural disasters due to climate change and abnormal weather, and the impact on ecosystems, the environment, agriculture, and other areas.

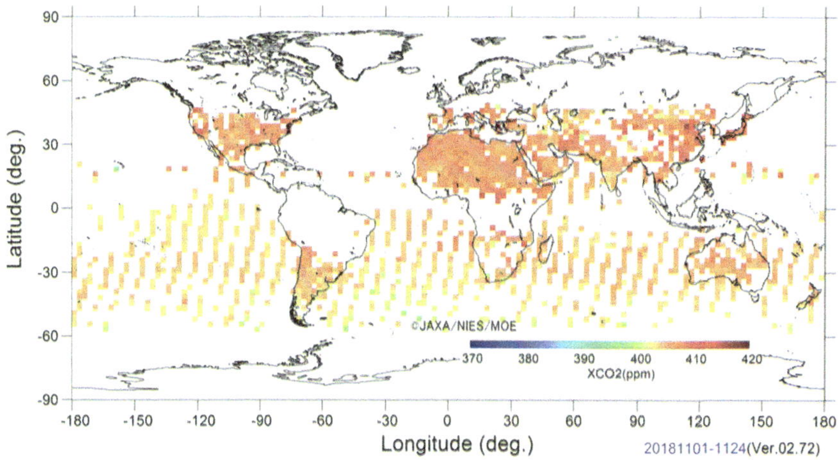

Fig. 13.2 Atmospheric carbon dioxide concentration observed by GCOM-C. (Exhibit: JAXA)

The Intergovernmental Panel on Climate Change (IPCC) warns that continued greenhouse gas emissions could lead to further warming and climate change. While a variety of activities ranging from international to individual energy-saving measures have been attracting attention to combat global warming, it is very difficult to quantitatively evaluate the results of these activities.

The Greenhouse Gases Observing Satellite "GOSAT" operated by JAXA can observe the concentration distribution of greenhouse gases such as carbon dioxide and methane from space (Fig. 13.2). Furthermore, the cloud and aerosol sensors can monitor air pollution of fine particulate matter (PM2.5, black carbon, etc.).

The sensor called the Multi-Wavelength Laser Imager (SGLI) onboard the Climate Change Observation Satellite "GCOM-C" operated by JAXA can measure the temperature of the Earth's surface in the wavelength range from visible light to infrared and thermal infrared, and acquire various data on climate change, such as global warming. This allows for the acquisition of various data on climate change, such as global warming.

As a global warming countermeasure, deforestation and forest degradation are observed by various satellite data. Carbon dioxide absorption by forests is estimated and used for global warming countermeasures. Satellite remote sensing technology is indispensable for understanding forests and evaluating policies to reduce greenhouse gas emissions on a global scale.

13.1.3 Climate Change

Climate is the change in atmospheric conditions over a long time period of several decades. When this climate changes over a long time scale due to various factors, it is called climate change. Climate change can be caused by natural factors, such as changes in solar activity and atmospheric changes due to volcanic eruptions, or by human factors, such as deforestation and increased greenhouse gas emissions due to fossil fuel consumption. Long-term monitoring is essential to minimize the effects of these changes.

Remote sensing by earth observation satellites can obtain global data that is difficult to observe on the ground. Since satellites repeatedly pass over the same upper-air orbit, the same locations can be observed over and over again, and changes due to seasons and years can be monitored over a long period of time.

Phenomena that are thought to be the effects of climate change are beginning to manifest themselves in a variety of ways. For example, climate change effects are projected to increase drought due to an increase in the number of rainfall-free days and a decrease in snow cover in some regions.

13.1.4 COVID-19

In 2020-2021, COVID-19 raged across the world. By using satellite remote sensing technology, we can capture the social changes caused by the COVID-19 pandemic. Although satellite data cannot directly observe the infection status of COVID-19, it can visualize the status of human and economic activities over a wide area.

The Sentinel-5 satellite, operated by the European Space Agency (ESA), can accurately observe ozone, methane, formaldehyde, aerosols, carbon monoxide, nitrogen dioxide, and sulfur dioxide in the air. After the outbreak of COVID-19, many factories stopped operating and the concentration of nitrogen dioxide in the atmosphere decreased. This was especially noticeable in China, where the clean air could be observed from space.

Satellites can also be used to observe the light of cities and towns at night. The light of a city at night can be evaluated as the level of activity in the city. Nighttime light can also be used to estimate the decrease in economic activity in a city due to an outbreak of COVID-19. Figure 13.3 shows changes in nighttime light around Hong Kong. On the left is March 2019 before the outbreak of COVID-19, and on the right is March 2020 before the outbreak of the COVID-19.

13.1.5 Forest Fires

Large-scale forest fires that are occurring all over the world are becoming a major problem. Forest fires not only cause loss of life and homes due to the spread of fire, but also affect many other areas, such as global warming due to increased carbon dioxide emissions, and the ecosystem and natural environment due to the burning of forests. There are two causes of forest fires: natural ignition and human-caused factors. Spontaneous fires are caused by lightning strikes, volcanic eruptions, and

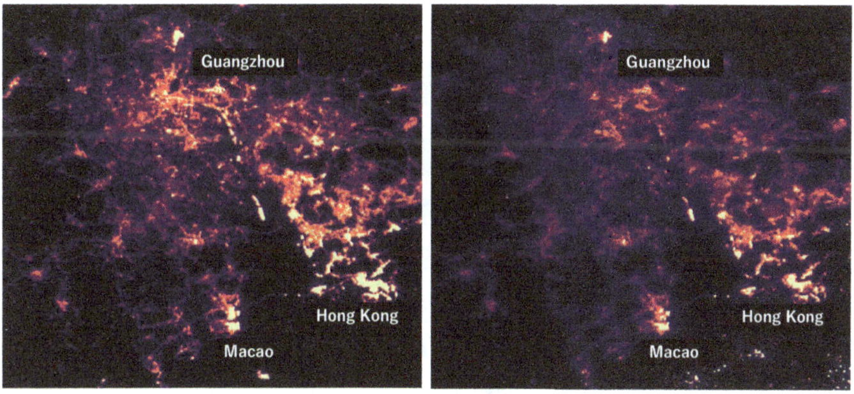

Fig. 13.3 Nighttime light before and after the spread of COVID-19

other factors. On the other hand, forest fires caused by human factors are caused by fires started by lighting bonfires, smoking cigarettes, or burning fields.

The Terra and Aqua satellites operated by NASA in the United States, which can observe forest fires twice a day, are effective methods for monitoring forest fires. The MODIS sensor on these satellites can observe wavelengths in the thermal infrared region and monitor ground surface temperature, fires, eruptions, sea surface temperature, etc.

13.1.6 Malaria

There is also growing sensitivity to the risk of mosquito-borne infectious diseases such as malaria and dengue fever, which are spreading in tropical and subtropical countries. Mosquitoes are said to be "the world's most prolific killer of humans," and it is still fresh in our minds that Japan reported a domestic outbreak of dengue fever several years ago, as well as an epidemic of mosquito-borne Zika fever.

In order to link local data on infection to measures at the national and regional levels, it is necessary to use environmental and demographic data at various scales and to share the findings obtained. By comparing field survey data on disease incidence with environmental conditions such as weather conditions, distribution of forests and water bodies, and land use, infection risk can be clarified from an areal perspective for the first time.

In order to deter the spread of infection, the distribution of residences, people's movements (from daily to seasonal movements, disaster evacuation, etc.), livelihood and residential activities, etc. must be ascertained, and focused and intensive inspection and isolation of persons at risk of infection are necessary (Fig. 13.4).

Remote sensing technology using satellite data can be used to construct land cover classification maps, forest distribution maps, monthly soil moisture maps, and

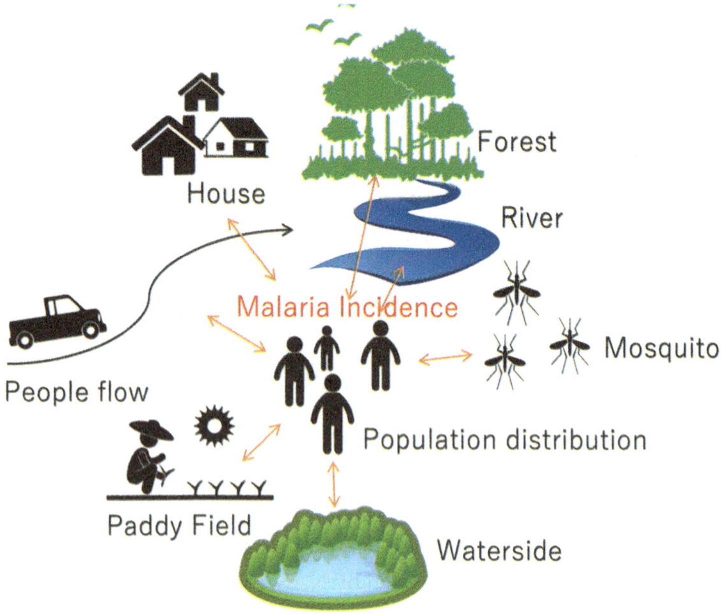

Fig. 13.4 Geospatial information related to malaria

distribution maps of paddy fields, etc. (including monthly waterlogging conditions, etc.). In addition, long-term archived satellite data is available free of charge, allowing monitoring of environmental change trends. These environmental maps provide important basic information for estimating the habitat areas of malaria-carrying mosquitoes.

These environmental maps can be integrated with medical records of medical institutions and patient distribution using GIS (Geographic Information System). By superimposing patient distribution, movement, and living pattern information, it will be possible to provide this information as infection warning information in response to changes in mosquito distribution potential areas due to environmental changes and other factors, as well as changes in people's living activity areas due to road infrastructure development and other factors.

13.2 International Cooperation in Times of Disaster

13.2.1 International Disaster Charter

Many large-scale disasters have occurred around the world, and satellite data is being used to assess damage and take countermeasures. The International Charter on Disaster Charter (Disaster Charter 2021) was established in 2000 as a framework for the international free provision of images from earth observation satellites. Since disasters occur in various locations and some of the affected countries do not have Earth observation satellites, the role of the International Disaster Charter is very important in disaster countermeasures.

The International Charter on Disaster Charter is a "Charter on Cooperation to Achieve Harmonized Utilization of Space Facilities in the Event of Natural or Technological Disasters" to which 17 organizations have subscribed and which is being used by a number of Earth observation satellites to conduct disaster observations (February 2021). Disasters covered by the International Charter on Disasters are defined as extremely difficult situations involving loss of life or major property damage.

JAXA has been participating since 2005, providing many satellite data from the Earth Observation Satellites "Daichi" and "Daichi-2". In providing data to the International Charter for Disaster Reduction, it is one of the most widely used satellite data for disasters. Since the International Charter for Disaster Charter activities started in 2000, until February 2021, there have been 699 activations, and the results of satellite data analysis have been provided to disaster-stricken agencies.

After the Great East Japan Earthquake in 2011, satellite data of more than 5000 scenes were provided from around the world through the International Disaster Charter to help assess the damage and assist in recovery and reconstruction. Figure 13.5 shows the tsunami damage along the coast of Miyagi Prefecture.

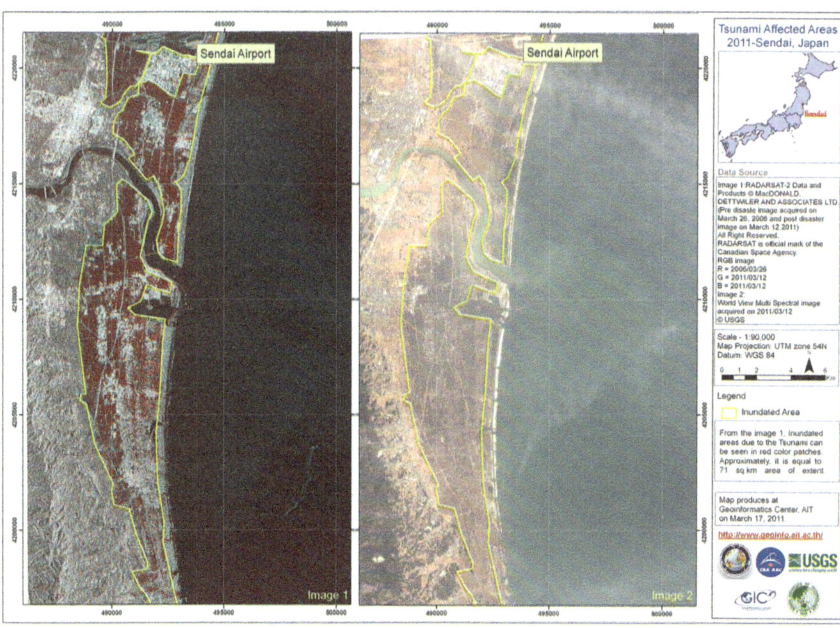

Fig. 13.5 Satellite data provided by the International Disaster Charter during the Great East Japan Earthquake (2011)

13.2.2 Sentinel Asia

According to statistics on natural disasters from 1986 to 2015 compiled by the Institute of Disaster Epidemiology at Leuven Catholic University in Belgium, the Asian region, including Japan, is the most affected by disasters worldwide. 39% of all disasters worldwide occur in Asia, which accounts for 61% of the world's total deaths and about 89% of the total number of people affected by disasters. In other words, for 39% of the world's total number of disasters, a great deal of the damage comes from Asia. In developing countries in particular, there are many areas where the social infrastructure on the ground is not well developed, and the lack of information is also a cause of the spread of damage. In such regions, it is essential to use satellite data for disaster response.

Looking at the use of satellite data during disasters in the Asia-Pacific region, only a limited number of countries have satellites that can monitor disasters, and many countries do not have national space agencies. Even in countries that do not have satellites, satellite data is an important source of information in times of disaster, which is why an international project called Sentinel Asia was launched in 2006.

Sentinel Asi (Sentinel Asia 2021) a was launched to support disaster management in the Asia-Pacific region using satellite remote sensing technology from space. Space agencies, disaster management agencies, international organizations, and universities in the Asia-Pacific region collaborate to conduct emergency observations using earth observation satellites and create disaster maps in response to requests from disaster management agencies in each country when a large-scale disaster occurs. As of February 2021, Sentinel Asia has 91 member organizations from 28 countries and regions, including Japan (Fig. 13.6).

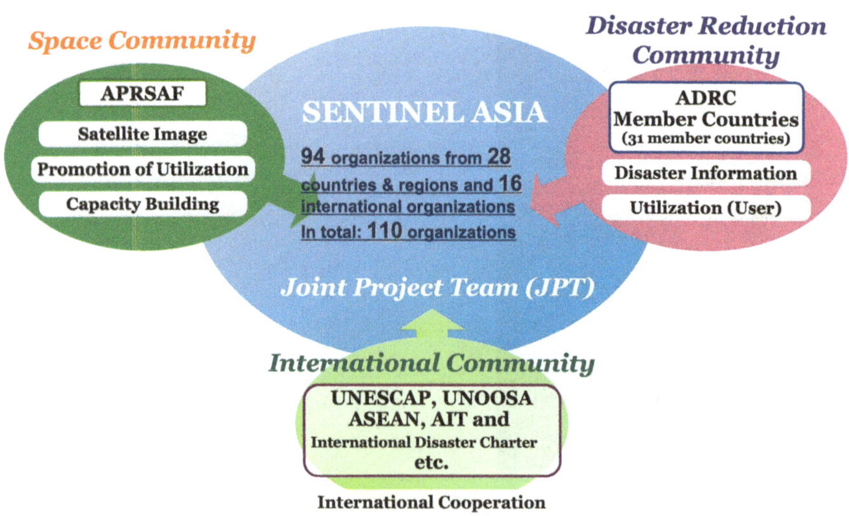

Fig. 13.6 Sentinel Asia member countries. (Exhibit: Sentinel Asia)

13.3 Summary

In this chapter, examples of the use of geospatial information and remote sensing for international cooperation are given, and efforts from a global perspective are explained. Global-scale issues such as the SDGs and global warming and international cooperation in times of disaster are explained.

References

Disaster Charter (2021) International Disaster Charter web site. https://disasterscharter.org

Sentinel Asia (2021) Sentinel Asia web site. https://sentinel-asia.org

UN (2021) United Nations web site. https://www.un.org/sustainabledevelopment/sustainable-development-goals/

Part IV
Open Data and Personal Data in the Use of Geospatial Information

Chapter 14
The Advent of a Participatory Data Society and the Usage of Open Data

Toshikazu Seto

Abstract The development of information and communications technology (ICT) has not only made it easier for us to obtain a vast amount of information in our daily lives, but also created a system in which we can become senders of information through the Web. In this section, we discuss how various types of geospatial information can be utilized in our social lives through participatory sharing methods.

Keywords Crowdsourcing · Participatory GIS · Volunteered geographic information · Citizen science · Open data

14.1 Data Sharing Through User Participation

14.1.1 Crowdsourcing

The term "crowdsourcing" was coined by Jeff Howe, one of the editors of Wired magazine, to describe the phenomenon of outsourcing by companies and other organizations to crowds scattered across the Internet since the mid-2000s (Howe 2008; Grier 2013). Although it is currently defined in various ways by commentators in many fields, its general definition is "the outsourcing of some kind of contribution, such as asking an unspecified number of people to perform a task or contribute money" (Morishima 2020). It can be broadly categorized into remunerative and non-remunerative types.

Crowdsourcing services that involve remuneration include business matching services such as "Crowdworks" and "Lancers." In these services, the client solicits individuals to undertake certain work, such as marketing research, design development, website creation, or system construction. The client negotiates and contracts with the applicants using e-mail, chat, online meeting tools, etc., and then pays a fee

T. Seto (✉)
Faculty of Letters, Komazawa University, Tokyo, Japan
e-mail: tosseto@komazawa-u.ac.jp

Y. Sekimoto, Y. Kawahara (eds.), *Utilization of Geospatial Information in Daily Life*, New Frontiers in Regional Science: Asian Perspectives 65,
https://doi.org/10.1007/978-981-19-6905-8_14

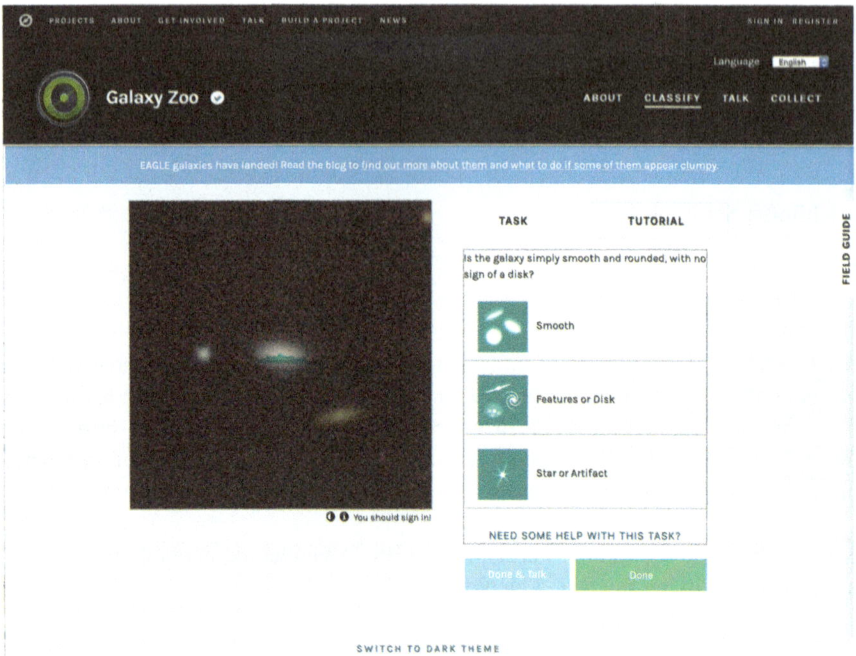

Fig. 14.1 Galaxy Zoo, a participatory image classification site for galaxy surveys. (Source: http://www.galaxyzoo.org/)

based on the results obtained. This payment system is based on the results obtained. Another such example which has been around for a while is "Amazon Mechanical Turk," which started in 2005. This is a crowdsourcing service that specializes in outsourcing more detailed work units, such as tagging photos and videos, data verification, and improving the accuracy of translation patterns. These services are also referred to as "micro tasking" because they target small-scale data that are difficult to process automatically by machines.

On the other hand, crowdsourcing is also being applied to non-remunerated research and scientific activities. A typical example is the Galaxy Zoo (Fig. 14.1). This is one of the earliest web-based crowdsourcing activities that began in 2007. More than 150,000 people have participated in this web-based categorization of galaxy image data taken by space telescopes. In the first year of the project, approximately one million galaxy data were created, and more than 50 million classifications were made. The classification data are open to the public through the Web and are basic data for space science that can be used for secondary purposes as open data, as described later in this chapter.

14.1.2 Participatory Mapping

Participatory mapping is an activity designed to map local geographic knowledge represented by community maps, based on people's proactive participation, while making full use of ICT. This kind of mapping practice is also closely related to "Participatory GIS" (Wakabayashi et al. 2017), which has been a subject of GIS research since the late 1990s. Specifically, it is not only urban activities such as community management and urban development, but also the use of maps and GIS in local resource management methods to improve empowerment (Friedman 1992) in areas where maps are not sufficiently developed or where the rights of indigenous people have been severely restricted. It has also attracted attention as a participatory mapping activity. Traditionally, such activities have been carried out using paper maps and three-dimensional topographical models. However, in recent years, even in mountainous areas and developing countries, satellite imagery and remote sensing data have been actively used in the field, and GPS receivers and other devices have become smaller and cheaper for positioning on the ground, allowing us to grasp the state of the land over a wide area.

Many of the activities that fall into this participatory mapping category have a specific theme or purpose and are often localized or limited by specific time periods. Safecast, for example, is a worldwide project to share radiation measurement data in the form of an open license that can be freely used; a grassroots sensor network has been established, including data collection through various sensors and homemade Geiger counters (Fig. 14.2). Activities such as Galaxy Zoo and Safecast, which

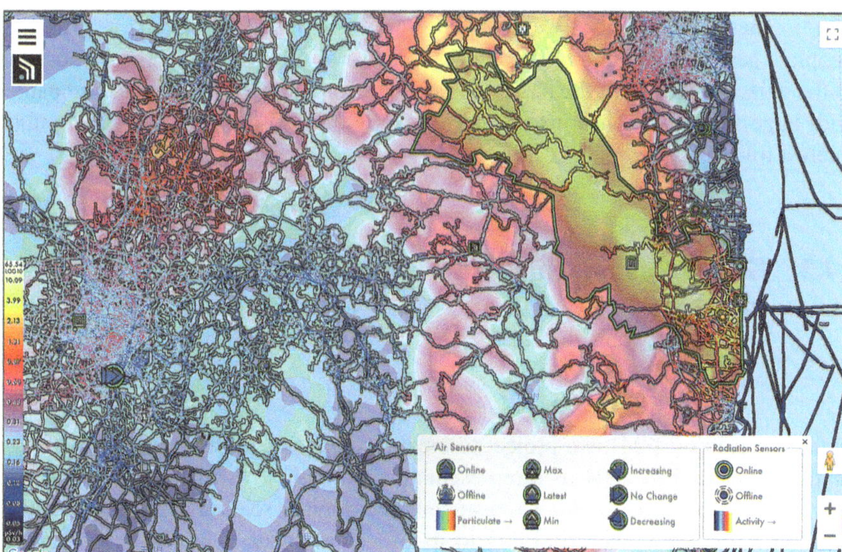

Fig. 14.2 Safecast, a website for sharing radiation levels measured by Geiger counters. (Source: https://map.safecast.org/)

include scientific elements and involve data collection with not only experts but also interested citizens, and which can lead to scientific research and, depending on the theme, policy reflection, are also known as "citizen science." As a place for co-creation between citizens and researchers, the 5th Science and Technology Basic Plan of Japan specifically calls for social understanding and infrastructure development to promote these activities (Young Academy of Japan 2020).

14.1.3 Volunteered Geographic Information

Geospatial information, mainly through maps, has often been created by the state, government agencies, or private companies. Therefore, it is not always easy for us to obtain and use this information, not only because of the frequency of updates, but also because of the price when the information is paid for. On the other hand, there is an increasing need to grasp the ever-changing geographical situation on Earth, and in many cases, such as in developing countries, there are no commercial maps produced by the state or government agencies. In today's world, where the Web is widely used, the geographic knowledge that forms the basis of maps is being shared spontaneously by non-specialists in addition to those who are positioned as experts in geography and regions, using crowdsourcing and participatory mapping methods described above.

Based on this social trend, Goodchild proposed "Volunteered Geographic Information (VGI)," which is a grassroots activity which reverses the flow of geospatial information that has traditionally been provided in a top-down manner (Goodchild 2007). These data have also become a vast information resource for exploring people's geographical knowledge and the nature of their living world. In addition to the citizens, who were the targets of similar efforts in the past, private companies with large-scale location information (log data) are also expected to play a public interest role in VGI.

14.2 Grassroots Sharing and Practice of Geospatial Information

14.2.1 OpenStreetMap

OpenStreetMap (OSM) is a project to create a database of "free" geospatial information. It started in 2004 as a representative example of VGI (Fig. 14.3). It was founded by Steve Coast, who was a student of computer science at the University of London at the time, and according to its official website, it has become a global project with about 7.5 million users as of February 2020.

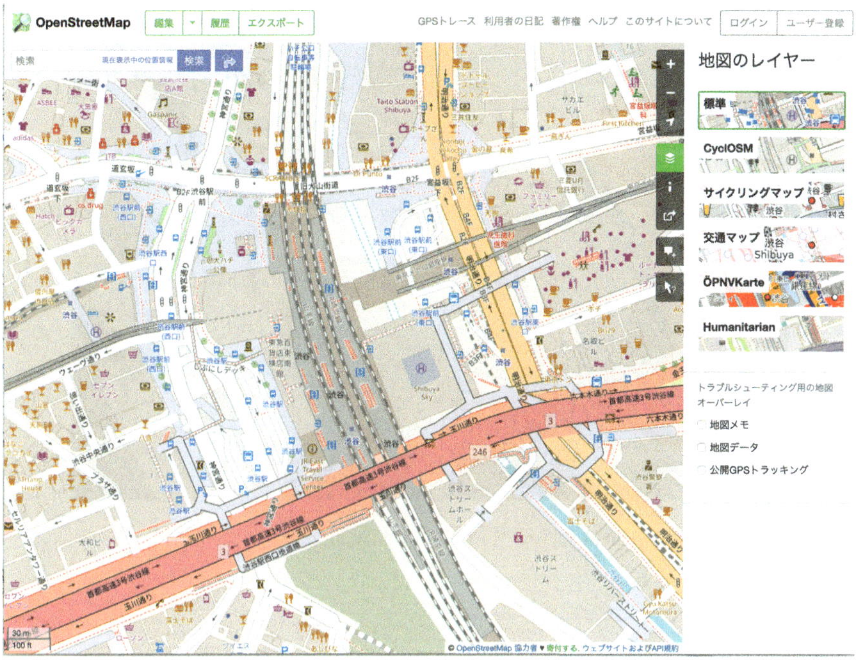

Fig. 14.3 OpenStreetMap, a free world mapping project (Shibuya area). (Source: http://www.openstreetmap.org/)

In OSM, geospatial information created by cartographic contributors (also known as mappers) is provided as archival data as well as Web maps through various map styles, all of which are open data that can be redistributed and used commercially. All of them are provided as open data (see Sect. 14.3) that can be redistributed and used commercially. Therefore, they are used not only as detailed road maps, including route information, but also as outdoor maps for mountain climbers and bicyclists, background, and route maps for car navigation systems, and in recent years, as digital data in some products designed for clothing and miscellaneous goods. However, because OSM is based on the principle of localism to ensure free licensing to the maximum extent possible, reprinting or data conversion (input) from Google Maps or other general web maps is not allowed in principle.

For this reason, activities to create and share geospatial information in OSM are often conducted through events called "mapping parties" in addition to individual activities using tools (Fig. 14.4). This is an event for mappers and new users to learn how to enjoy OSM, its basic rules, and data creation methods, while having fun together, and has been held mainly by the OSM community in various parts of Japan. Most mapping parties are based on maps using satellite imagery available in OSM, and additional information is often input directly on paper or with digital devices based on visual surveys in the field. One of the external services using OSM data is a service that can output paper maps at will (Fig. 14.5) and there are also applications for smartphones that support simple data input.

Fig. 14.4 Mapping party (field survey) for OSM data creation

Fig. 14.5 Field papers, a paper map-making service for OSM data creation. (Source: http://fieldpapers.org/)

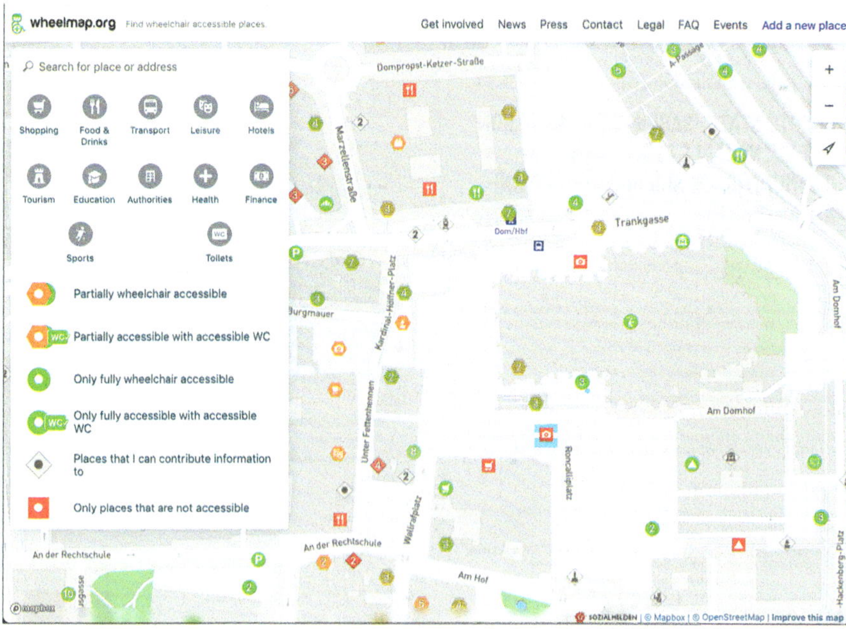

Fig. 14.6 Wheelmap, a web map for wheelchair users. (Source: http://wheelmap.org)

In the UK and European Union (EU) countries, where OSM activities are active, mapping parties are held every week in public places, mainly in urban areas, such as community facilities, universities, and even pubs. In Japan, similar activities have become increasingly popular in recent years as OSM has become increasingly active. In mapping parties, fieldwork is conducted to find geospatial information for OSM, which has the effect of revealing the parts of the city not depicted on existing maps, thereby changing the way participants see the city. In recent years, a project called "Wheelmap," which shares information specifically for the convenience of wheelchair users, has been actively conducted as a derivative project linked to OSM (Fig. 14.6).

14.2.2 Crisis Mapping Activity

Since OSM widely targets the creation of world maps, it mainly focuses on the sharing of basic and permanently existing geographic features such as roads, buildings, and land use (therefore excluding certain temporary features such as temporary facilities set up during crisis events). However, in the aftermath of the January 2010 earthquake in Haiti, a humanitarian aid initiative called crisis mapping was launched as a global effort to accurately map emergencies such as the immediate aftermath of a disaster (Meier 2015). In particular, the Haiti earthquake in March 2011 was a major cause for concern. In particular, the Great East Japan Earthquake that occurred

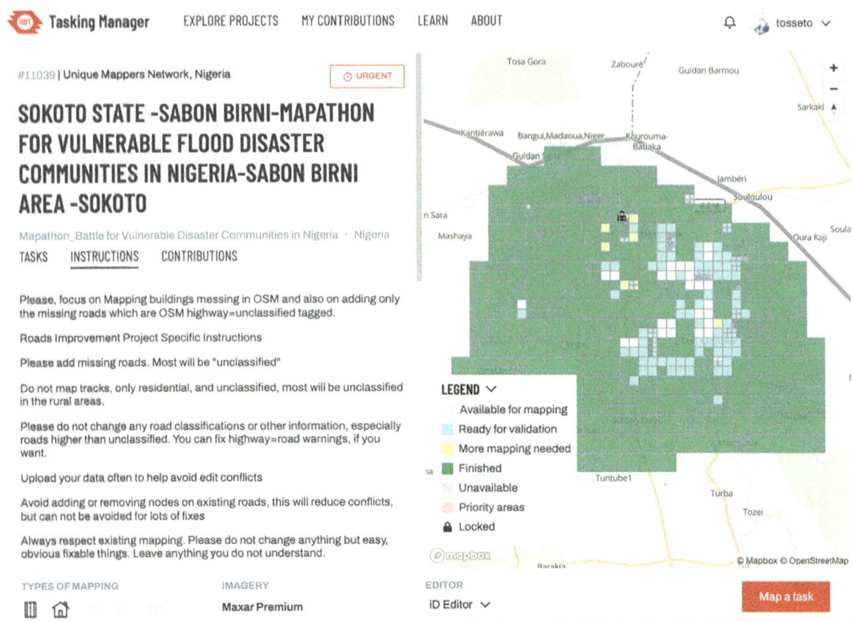

Fig. 14.7 Tasking Manager, a task management tool for crisis mapping using OSM (Vulnerable flood disaster mapping in Nigeria). (Source: https://tasks.hotosm.org/)

in March 2011 was the first time that large-scale crisis mapping was conducted in Asia. OSM data entry of damage using satellite imagery specially provided at the time of the disaster was used and aerial photographs taken in an emergency were actively conducted. In Japan, it is noteworthy that sinsai.info, introduced in Chap. 10, was immediately set up by volunteers skilled in ICT. However, it is now closed.

Thus, crisis mapping can be positioned as an activity to create and update local infrastructure maps that reflect the damage caused by a disaster, as typified by OSM. In addition, it is an activity to map information from social networking sites and various organizations through the Web, which is then integrated, shared, and analyzed for use in disaster response. The Humanitarian OpenStreetMap Team (HOT), which was established as an offshoot of OSM, is responsible for mapping all aspects of humanitarian assistance, including assessing damage from natural disasters and mapping urban infrastructure related to infectious diseases such as Ebola and COVID-19. In particular, HOT activities are being used in all aspects of humanitarian assistance.

HOT is also responsible for basic mapping activities using OSM, and by collaborating with international organizations, HOT is building a data model for collecting geospatial information useful for disaster response and developing field survey tools and operation guides that comply with this model (Fig. 14.7). Many of these are published on the Web and provided as licenses that can be freely reused, making them useful as tools for disaster prevention and mitigation.

14.2.3 Mapillary: Web-Based Service for Sharing Street Landscape Images

Mapillary, a project for sharing landscape images similar to OSM, has become a major collaborative sharing activity of street photos in recent years. Mapillary, which is a Swedish venture company that was founded in 2013 (although it was acquired by Facebook in June 2020, the project itself continues), is a platform for sharing landscape images, mainly on a street-by-street basis, and is closely linked to the activities of OSM. It works on the open data (Creative Commons Attribution Share Alike 4.0 [CC-BY-SA 4.0] in principle) of landscape images and the detection of land objects and objects using advanced artificial intelligence (AI) technology.

As of February 2021, more than 1.41 billion images have been shared in 190 countries and regions, and together with GPS data, they can be used as reference data for map editing on OSM through application programming interfaces (APIs). In addition, the 25,000-image Vistas Dataset (Fig. 14.8), which has labels for more than 100 types of objects for machine learning, is available for researchers, and landscape photos from around the world collected by grassroots users can be used as basic data for developing technologies for automated driving and improving machine learning algorithms.

14.3 Open Data and Geospatial Information

14.3.1 Open Data and Utilization for Geospatial Information

The sharing of geospatial information, mainly at the grassroots level, is spreading on a global scale through a variety of methods, many of which include licenses that allow redistribution and reuse. This trend is also spreading to national and local governments and other public organizations. Open data is a concept symbolizing this

Fig. 14.8 An example of the Mapillary Vistas Dataset (left: actual landscape images, right: images classified for machine learning). (Source: https://www.mapillary.com/dataset/vistas)

trend. According to Open Knowledge Foundation (2015), it is defined as "data that can be freely used and reused, and that can be redistributed by anyone." Open data was triggered by Data.gov, which was developed by the first Obama administration in the United States in 2009, and administrative information is now being opened using portal sites in Europe and Japan. In addition to the free use of these data, it is recommended that the data be released in its original form before processing and aggregation and that it be maintained in a machine-readable format.

One notable development related to geospatial information and open data is the signing of the Open Data Charter at the 39th G8 Summit in June 2013. The Charter emphasized the need for transparency to share the activities of public institutions and the realities of society with citizens, as well as the basic openness of data to be created by government institutions in the future. Geospatial information, mainly in the form of maps, was selected as a high-value data set to promote openness, along with transportation and infrastructure data, and has become a global concern (Sui 2014). In recent years, many application development events called "Ideathons" and "Hackathons" have been held in Japan, the United States, and Europe, as platforms for citizens to participate in open data practices. These events have been emphasized to motivate citizens to participate in the utilization of open data.

According to a survey by the Information Technology (IT) Strategy Office of the Cabinet Secretariat of Japan, the number of local governments in Japan that have made open data available to the public reached 915 as of December 2020 (approximately 51% of the total), with major cities with populations ranging from 200,000 or more to government-designated cities achieving a rate of 95% or higher. In the first half of the 2010s, when the open data policy began in earnest in Japan, the use of large-scale geospatial information as open data was limited to precedents such as the Shizuoka Prefecture, Muroran City, and Sabae City. However, as the government has been working on guidelines, the recommended data sets include the location information of various public facilities, borehole maps, and basic urban planning survey information.

14.3.2 Utilization of Data Through Citizen Collaboration for Open Government

The goal of open data is not only the free use of administrative information, but also establishing an open government, by methods such as improving the efficiency of administrative operations and citizen collaboration through the release of data (Goldstein and Dyson 2013). The mechanisms for collaboration between citizens, the private sector, and local governments through open data have been introduced in the government Chief Information Officers' portal (CIO portal) "Open Data 100" (https://cio.go.jp/opendata100), which can be used as a reference for activities.

For example, the Sapporo Nursery School Map (Fig. 14.9) was launched by the Code for Sapporo in October 2014 and allows users to search for nursery schools and

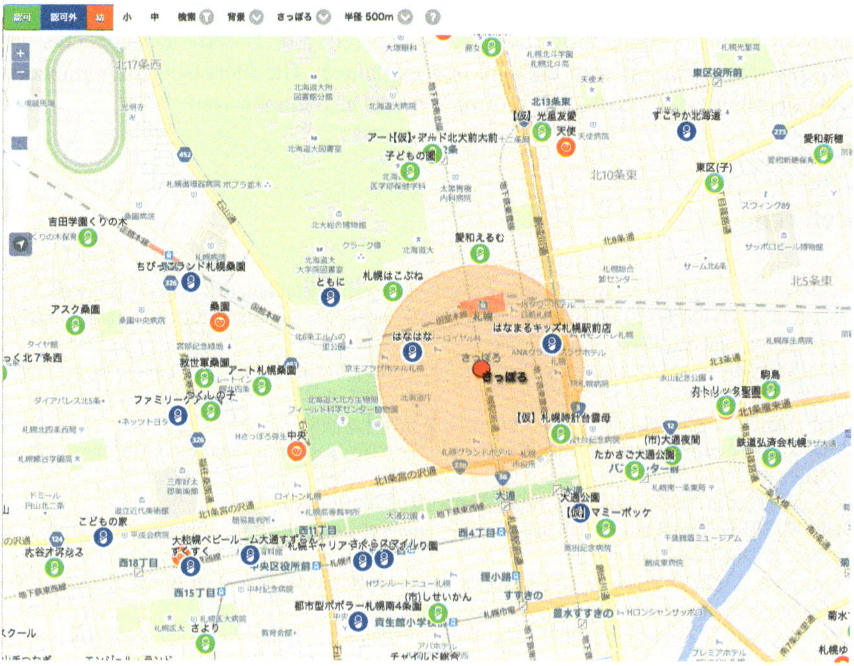

Fig. 14.9 Map of Sapporo Nursery Schools. (Source: http://papamama.codeforsapporo.org/)

kindergartens in Sapporo City on a map. In the past, nursery schools and kindergartens were under different administrative jurisdictions, and there was no centralized information, making it difficult to collect and organize information from various websites. This website is expected to reduce the burden on parents and guardians, as it allows them to check not only the location of nursery schools, but also their opening hours and availability on a map. Another characteristic of this open activity is that the structure and data format of this site has been released as an open source, and similar sites using the same structure have been developed in more than 12 regions to date.

As the citizen side of such open government activities, an activity which has been growing in recent years is "Civic Tech" (Inatsugu 2018), in which citizens use information technology and data to solve local issues. In Japan, there are more than 80 activities by region and theme, including the Code for Japan, which supports Civic Tech activities nationwide, and the Code for Kanazawa.

14.4 Conclusions

This chapter, along with the global deployment of ICT, demonstrated how many people are engaged in participatory activities to share geospatial information using the Internet, especially focusing on the social practice examples of grassroots communities. The current trend of open data and open government as activities to improve the operational efficiency and transparency of government agencies and facilitate collaboration with citizens was explained. The increasing importance of open distribution of geospatial information was also emphasized.

References

Friedman J (1992) Empowerment: the politics of alternative development. Wiley-Blackwell

Goldstein B, Dyson L (2013) Beyond transparency: open data and the future of civic innovation. Code for America Press, San Francisco

Goodchild MF (2007) Citizens as sensors: the world of volunteered geography. Geo J 69(4): 211–221

Grier DA (2013) Crowdsourcing for dummies. Wiley, London

Howe J (2008) Crowdsourcing: why the power of the crowd is driving the future of business. Crown Business, New York

Inatsugu H (ed) (2018) Civic tech. Keiso Shobo, Tokyo. (in Japanese)

Meier P (2015) Digital humanitarians: how big data is changing the face of humanitarian response. CRC Press, Boca Raton, FL

Morishima A (2020) Science and methods for crowdsourcing (Kyoritsu Smart Selection 32). Kyoritsu Shuppan, Tokyo. (in Japanese)

Open Knowledge Foundation (2015) The open data handbook. https://opendatahandbook.org/

Sui D (2014) Opportunities and impediments for open GIS. Trans GIS 18(1):1–24

Wakabayashi Y, Imai O, Seto T, Nishimura Y (2017) Theory and applications of participatory GIS. Kokon Shoin, Tokyo. (in Japanese)

Young Academy of Japan (2020) Recommendations toward the construction of a social system that promotes citizen science. Science Council of Japan, Tokyo. (in Japanese)

Chapter 15
Future Geospatial Information in Harmony with Advanced Technology and Human Life

Yoshihide Sekimoto

Abstract In the last part of this lecture, I would like to give an overview of the future of geospatial information in our daily lives and how it can be used in harmony with our daily lives. In particular, while there are aspects of advanced measurement technology and artificial intelligence, I would like to take a bird's eye view of how we can dispel the sense of anxiety from the standpoint of a citizen and how we can work together.

Keywords Measurement technology · Personal information · Artificial intelligence · Citizen collaboration

15.1 Various Advanced Measurement Technologies

15.1.1 Sophistication of Measurement and Control from the Air

As I have already mentioned in several chapters, in the field of measurement and control from the sky, ultra-high resolution remote sensing technology and UAV (Unmanned Air Vehicle) technology, which controls its own flight, have recently emerged and are becoming more and more accurate in both time and space. For example, in the former case, Google acquired Skybox in June 2014 and has been providing images from its commercial Earth observation satellite SkySat (0.85 m visible/near-infrared resolution) free of charge to nonprofit organizations since October (Ascii n.d.). In addition, the Advanced Land Observing Satellite-2 (ALOS-2, Daichi-2), launched by Japan in May 2014, is equipped with the Synthetic Aperture Radar (PALSAR-2) and will provide global 5-m resolution elevation (DEM) data in 2015.

Y. Sekimoto (✉)
Center for Spatial Information Science, University of Tokyo, Tokyo, Japan
e-mail: sekimoto@iis.u-tokyo.ac.jp

Y. Sekimoto, Y. Kawahara (eds.), *Utilization of Geospatial Information in Daily Life*, New Frontiers in Regional Science: Asian Perspectives 65,
https://doi.org/10.1007/978-981-19-6905-8_15

In addition, in the latter UAV, Amazon announced a delivery service called "Amazon Prime Air" using small unmanned aircraft for deliveries, which will start as early as 2015. The company is awaiting approval from the U.S. Federal Aviation Administration (FAA), but says it aims to deliver within 30 min of receiving an order (itmedia n.d.).

15.1.2 Advancement of Ground-Based Measurement and Control

On the other hand, of course, measurement and control from the ground is also advancing. In terms of cars on the road, it has been several years since the state of California signed a bill allowing Google to drive automatically on public roads in September 2012. The impact has been huge, and automated driving is being promoted in various parts of the world, with Japan in particular conducting research on driving in convoys, etc. In February 2013, the New Energy and Industrial Technology Development Organization (NEDO) announced that it had succeeded in having trucks drive at 4-m intervals on a test course. In February 2013, NEDO announced that it had succeeded in having trucks run at 4 m intervals on a test course (Nedo n. d.). In terms of the legal system, driving a car without touching both the steering wheel and gas pedal is legally prohibited, but in 2012, Google's robot car received a license in the state of Nevada. Also, in Japan, as part of the government's current deregulation policy, Tsukuba City was approved as a "Special Zone for Mobility Robot Experiments" in March 2011, and Segways and robots are allowed to drive on public roads in an area centered on the urban area of Tsukuba (Council.rt n.d.).

Finally, automation is also progressing in construction sites and farms. For example, in the past, rice had to be grown into seedlings before planting due to developmental problems, but with automation, rice can be processed with steel so that it can be directly spread. However, the direct sowing method, which combines the use of iron and other processing methods to directly sow rice, has greatly improved the efficiency of the operation (Nakatani 2014).

15.1.3 Measurement of People's Behavior

People's behavior is also being measured at a finer level of quality, not simply by GPS and accelerometers to determine position and direction. Here, I would like to introduce the case of Hitachi, Ltd.'s "Business Microscope," which was also featured in the *Harvard Business Review* in February 2015 as a "historic wearable device." In this system, when two employees wearing nametag-type sensors with multiple built-in sensor devices approach each other within a certain distance, they communicate with each other to detect that they have met face-to-face, while storing

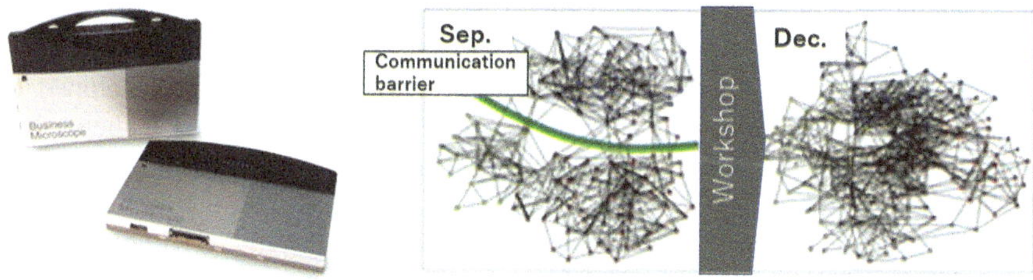

Fig. 15.1 Business microscope. (Provided by Hitachi, Ltd. Left: nametag-type sensor with multiple built-in sensors; Right: monitoring changes in communication in the workplace using nametag-type sensor)

their face-to-face time and physical activity rhythms on a server to monitor communication and activity within the company (Fig. 15.1, left). In fact, in the right part of Fig. 15.1, we can clearly see how communication in a certain workplace was divided into two parts, but after the workshop was held, communication progressed. There will be more cases where this kind of human-centered approach will be used to improve the activities of workplaces and other groups.

15.2 Diffusion of Artificial Intelligence Technology and Beyond

15.2.1 Breakthroughs in Deep Learning and the Importance of Data

You have probably heard the term artificial intelligence (AI) before. The important point is that it learns something from the data it is given, and is able to make its own estimates. In simple terms, it can take an image and automatically determine if it is an apple or a person, and more recently, you may have heard that Google's Alpha Go program defeated the world's top Go masters without handicap in 2015. This is another example of a computer learning from many past game records and coming up with patterns that even the masters could not come up with, and in turn, creating new patterns in the world of Go.

When learning something from data, defining parameters and features of some kind of model and estimating them is called machine learning, and it is a concept that has been around for a long time. However, what the feature values should be depends on the characteristics of the target, so it had to be defined in an ad-hoc manner. On the other hand, neural networks have been developed to map the correspondence between the given data and the resulting labels (for example, information such as apples or people, or how the next move was made in Go) with as much accuracy as possible. However, since it mechanically estimates the weights of each neuron, it tends to over-learn the given data, and there have been cases where it could not estimate well when new data is given.

In such a situation, multilayer neural networks (deep learning) developed by Hinton et al. in the 2000s and later succeeded in learning the features themselves well, ensuring versatility, greatly improving accuracy, and becoming widely used at the business level.

However, it is also true that these methods cannot do anything without data. In general, it is said that several hundred to several thousand pieces of data are needed to obtain some degree of accuracy for estimation. For example, the recognition of road space familiar to our daily life, especially road damages such as potholes, cracks, and tortoise-shaped cracks, is important for cars, and will be even more important in the future when automatic driving becomes widespread, but even for such things, it is difficult to obtain data. In such a situation, Fig. 15.2 shows how the

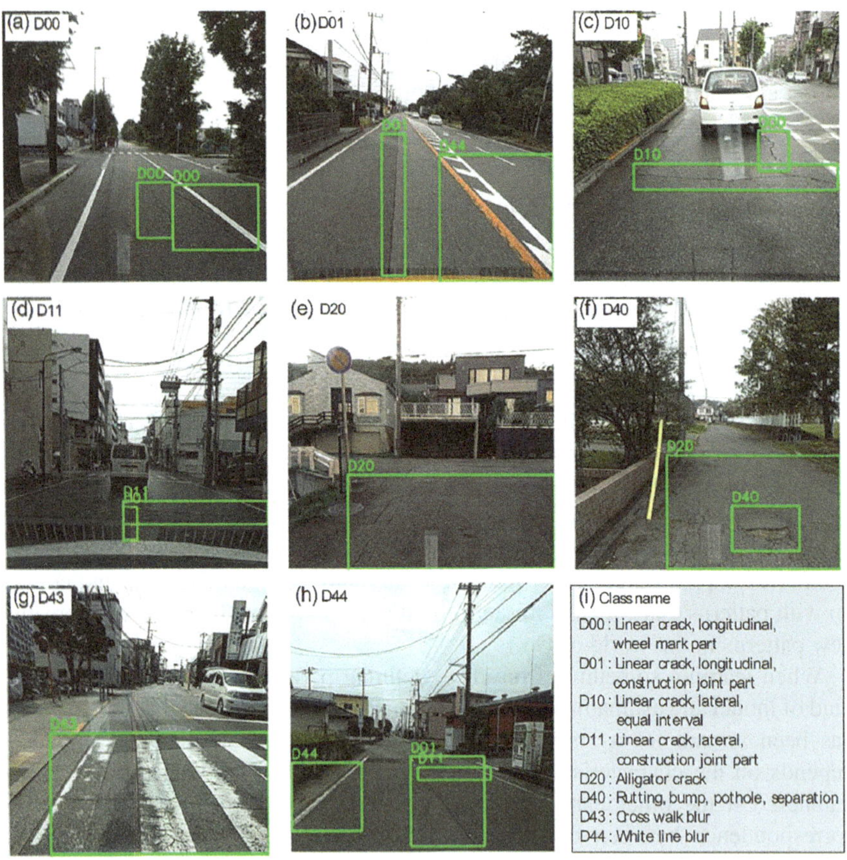

Fig. 15.2 Road damage detection from smartphones using deep learning. (Cited from Maeda et al. 2018)

authors, in collaboration with several local governments in Japan, collected road images through road management vehicles, labeled about 10,000 road damage images as shown above, and published them on the Github website in 2018 (see Maeda et al. 2018 for details). These have led to its use by researchers in many countries.

15.2.2 Pseudo Data Generation (Fake Data) Using GAN

However, as the accuracy of estimation increases in the form of deep learning, it becomes possible to generate similar images. Have you ever heard of the term "fake data"? Fake data is an image that looks like a person's own image, even though it is not an image of the person actually doing something. This kind of technology is called GAN (Generative Adversarial Network). Of course, stories such as fake data may lead to misuse, so ethical measures are necessary. However, on the other hand, it is important to consider using it in a positive direction.

For example, in Sect. 15.2.1, we talked about images of road damage. In actual road management, we want to know about damage such as potholes as soon as possible because it may lead to accidents, but it is surprisingly difficult for a third party to find images of potholes because they quickly return to their original state after repair. In such a case, there is a possibility to use existing collected images that have been well processed by GAN. Figure 15.3 shows how much of the image is mixed. It can be seen that the best accuracy is obtained when the GAN-generated similar images are not mixed at all (0%), 50% or 100% generated and mixed (50 or 100%) with the 1200/800/400 pothole images that we actually have. It has been found that a moderate level of generation is better than no mixing at all or too much mixing.

		Available data (images)		
		1,200	**800**	**400**
Additional data	0%	0.39	0.33	0.32
	50%	**0.41**	**0.37**	**0.37**
	100%	0.28	0.34	0.34

Fig. 15.3 Estimation accuracy (F-values) when using a mixture of data generated pseudo by GAN as teacher data. (Cited from Maeda et al. 2020)

15.2.3 Toward a Truly Collaborative Society that Raises the Awareness of Citizens

Finally, some people may feel that with the spread of artificial intelligence, everything will become automatic, and they will be afraid of being controlled by robots. There is also the term "singularity" (technological singularity), which was coined by Ray Kurzweil in his 2005 book that artificial intelligence with a will may replace humanity as the main driver of civilization's progress around the year 2045. I won't go into this in depth here, but how should we deal with artificial intelligence in our daily lives?

Figure 15.4 shows a report submitted by citizens on problems in the city through a citizen collaboration report application called "Chiba Repo," which was launched by Chiba City in August 2014. This is mapped on the public website, but when citizens have complaints about the city, they often call or go to the city office. However, it is surprisingly difficult to accurately report the status and location of problems over the phone or at the counter, and it takes time for both the citizen and the city hall staff to identify the problem, which is stressful for both parties. On the other hand, if a citizen sends an app with an image, location information by GPS, etc., and a short message, the staff member who receives the message can easily understand it and can respond more calmly after discussing how to deal with it than by phone. As described in detail in Chap. 14, this kind of citizen collaboration is an effective tool

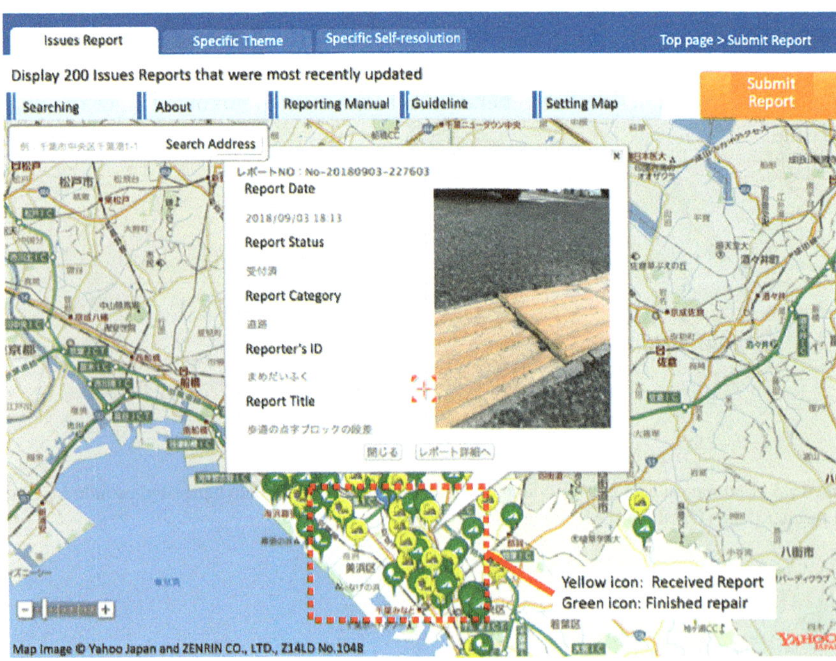

Fig. 15.4 Images submitted by Chiba Repo. (Cited from Seto and Sekimoto 2019)

for connecting citizens and the city in the IT age, and since fiscal 2019, it has been nationwide in the form of My City Report, which has spread to 11 municipalities, and is expected to increase further in the future.

The important thing about this My City Report system is that not only citizens can contribute, but, for example, road management staff can also place their smartphones on the dashboard of their patrol vehicles and use an app for road managers to automatically detect road damage using deep learning based on images acquired while driving. This is based on the technology described in Sects. 15.2.1 and 15.2.2, but if we only ask citizens to submit information about problems in the city, it may be thought that we are covering the work that city hall should be doing by exploiting the labor of citizens, and the citizens may not think that it is a collaborative effort. In this sense, road managers should also promote efficiency and IT in their daily patrols, and this is a good example of fostering the awareness of citizen cooperation.

15.3 For the Betterment of Society

In this text, Chaps. 2–5 have been about basic geospatial information technologies, Chaps. 6–13 have been about technologies for using geospatial information in individual fields, and Chap. 14 and this chapter have been about the future.

How much progress will be made and how widespread will the technologies related to geospatial information be in 2030, 10 years from now, or in 2040, 20 years from now? On the other hand, what social issues will be solved, and what social problems are yet to be solved or will emerge? I would like you to think about this carefully.

15.4 Conclusion

In this chapter, as a final summary, I have introduced the advanced measurement technology of geospatial information, and at the same time, I have looked at how it can contribute to the challenges of society and how it can alleviate the sense of anxiety about the future in the midst of increasing automation technologies such as artificial intelligence, and I have drawn a vision of the future of geospatial information in which advanced technology and human life are in harmony.

References

Ascii (n.d.). http://ascii.jp/elem/000/000/947/947865/
Council.rt (n.d.) Robot special zone demonstration experiment promotion council. http://council.rt-tsukuba.jp/

itmedia (n.d.). http://www.itmedia.co.jp/news/articles/1312/02/news056.html
Maeda H, Sekimoto Y, Seto T, Kashiyama T, Omata H (2018) Road damage detection and classification using deep neural networks with smartphone images. Comput Aided Civ Inf Eng 33(12):1127–1141. Available 30 June 2018
Maeda H, Sekimoto Y, Seto T, Kashiyama T, Omata H (2020) Generative adversarial network for road damage detection. Comput Aided Civ Inf Eng 36:47–60. Available 2 June 2 2020
Nakatani Y (2014) Trends of new technologies in agricultural machinery and expectations for fluid power, fluid power. 28(4)
Nedo (n.d.) Nedo news release, successful automated driving and convoy driving experiments of heavy trucks—results of energy project released, 2013. Available http://www.nedo.go.jp/news/press/AA5_100178.html
Seto T, Sekimoto Y (2019) Trends in citizen-generated and coordinated urban infrastructure feedback data: towards citizen-driven infrastructure management in Japan. ISPRS Int J Geo-Inf 8(3)

The manufacturer's authorised representative in the EU is Springer Nature Customer Service Centre GmbH, Europaplatz 3, 69115 Heidelberg, Germany. If you have any concerns regarding our products, please contact ProductSafety@springernature.com

Printed and bound by CPI Group (UK) Ltd, Croydon, CR0 4YY
30/09/2026
02225303-0001